About this book

This book is designed to help you get your best possible grade in your Mechanics 1 examination. The author is a Chief examiner, and has a good understanding of AQA's requirements.

Revise for Mechanics 1 covers the key topics that are tested in the Mechanics 1 exam paper. You can use this book to help you revise at the end of your course, or you can use it throughout your course alongside the course textbook, *Advancing Maths for AQA AS & A level Mechanics 1*, which provides complete coverage of the syllabus.

Helping you prepare for your exam

To help you prepare, each topic offers you:

- **Key points to remember** – summarise the mechanical ideas you need to know and be able to use.

- **Worked examples** – help you understand and remember important methods, and show you how to set out your answers clearly.

- **Revision exercises** – help you practise using these important methods to solve problems. Exam-level questions are included so you can be sure you are reaching the right standard, and answers are given at the back of the book so you can assess your progress.

- **Test Yourself questions** – help you see where you need extra revision and practice. If you do need extra help, they show you where to look in the *Advancing Maths for AQA AS & A level Mechanics 1* textbook and which example to refer to in this book.

Exam practice and advice on revising

Examination style paper – this paper at the end of the book provides a set of questions of examination standard. It gives you an opportunity to practise taking a complete exam before you meet the real thing. The answers are given at the back of the book.

How to revise – for advice on revising before the exam, read the **How to revise** section on this page.

How to revise using this book

Making the best use of your revision time

The topics in this book have been arranged in a logical sequence so you can work your way through them from beginning to end. But **how** you work on them depends on how much time there is between now and your examination.

If you have plenty of time before the exam then you can **work through each topic in turn**, covering the key points and worked examples before doing the revision exercises and Test Yourself questions.

If you are short of time then you can **work through the Test Yourself sections** first, to help you see which topics you need to do further work on.

However much time you have to revise, make sure you break your revision into short blocks of about 40 minutes, separated by five- or ten-minute breaks. Nobody can study effectively for hours without a break.

Using the Test Yourself sections

Each Test Yourself section provides a set of key questions. Try
each question:

- If you can do it and get the correct answer, then move on to
  the next topic. Come back to this topic later to consolidate
  your knowledge and understanding by working through the
  key points, worked examples and revision exercises.

- If you cannot do the question, or get an incorrect answer or part
  answer, then work through the key points, worked examples
  and revision exercises before trying the Test Yourself questions
  again. If you need more help, the cross-references beside each
  Test Yourself question show you where to find relevant
  information in the *Advancing Maths for AQA AS & A level Mechanics
  1* textbook and which example in *Revise for M1* to refer to.

Reviewing the key points

Most of the key points are straightforward ideas that you can
learn: try to understand each one. Imagine explaining each idea to
a friend in your own words, and say it out loud as you do so. This
is a better way of making the ideas stick than just reading them
silently from the page.

As you work through the book, remember to go back over key
points from earlier topics at least once a week. This will help you
to remember them in the exam.

CHAPTER 1

Kinematics in one dimension

Key points to remember

1 The area under a velocity–time graph gives the displacement.

2 The gradient of a velocity–time graph gives the acceleration.

3 The gradient of a displacement–time graph gives the velocity.

4 Average speed = $\dfrac{\text{Total distance travelled}}{\text{Time taken}}$

5 Average velocity = $\dfrac{\text{Change in displacement}}{\text{Time taken}}$

6 You need to learn these formulae.

  (i) $v = u + at$

  (ii) $s = ut + \dfrac{1}{2}at^2$

  (iii) $s = \dfrac{1}{2}(u + v)t$

  (iv) $v^2 = u^2 + 2as$

Note that:
u = Initial velocity
v = Final velocity
a = Acceleration
s = Displacement
t = Time

Worked example 1

The graph below shows a velocity–time graph for a train that is moving on a straight track in a goods yard.

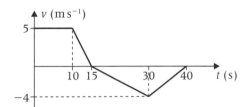

(a) Find the distance travelled by the train in the first 15 seconds.

(b) Find the total distance travelled by the train.

(c) Find the average speed of the train.

(d) Find the average velocity of the train.

(e) Find the acceleration of the train for $10 < t < 15$.

(a) This is given by the area under the curve between $t = 0$ and $t = 15$. In this case this has been done by splitting the area into a rectangle and a triangle.

$$\text{Distance} = 5 \times 10 + \frac{1}{2} \times 5 \times 5$$

$$= 50 + 12.5$$

$$= 62.6 \text{ m}$$

(b) To find the total distance the area of the triangle below the time axis is also required.

$$\text{Distance} = \frac{1}{2} \times 25 \times 4$$

$$= 50 \text{ m}$$

This can be added to the distance for the first 15 seconds to give the total distance.

Total distance = $62.5 + 50 = 112.5$ m

(c) The average speed is found using the total distance and the total time.

$$\text{Average speed} = \frac{\text{Total distance travelled}}{\text{Time taken}}$$

Using **4**

$$= \frac{112.5}{40}$$

$$= 2.81 \text{ m s}^{-1} \text{ (to 3 s.f.)}$$

(d) To find the average velocity the change in displacement is needed. The initial displacement is 0 and the final displacement is $62.5 - 50 = 12.5$.

$$\text{Average velocity} = \frac{\text{Change in displacement}}{\text{Time taken}}$$

Using **5**

$$= \frac{12.5}{40}$$

$$= 0.313 \text{ m s}^{-1} \text{ (to 3 s.f.)}$$

(e) The acceleration will be given by the gradient of the line on the graph.

$$\text{Acceleration} = \frac{-5}{5} = -1 \text{ m s}^{-2}$$

Using **2**

Worked example 2

A car is travelling at 30 m s^{-1}. The driver applies the brakes until the car stops. Assume that the car experiences a constant acceleration and travels along a straight line as it slows down. After the brakes have been applied for 2 seconds the car has travelled 50 metres.

(a) Find the acceleration of the car.

(b) Find the total time that it would take the car to stop.

(c) Find the total distance travelled by the car as it stops.

(a) Use the equation $s = ut + \dfrac{1}{2}at^2$ with $s = 50$, $u = 30$ and $t = 2$.

$$50 = 30 \times 2 + \tfrac{1}{2} \times a \times 2^2$$

$$50 = 60 + 2a$$

$$a = \frac{50 - 60}{2} = -5 \text{ m s}^{-2}$$

Using **6** (ii)

(b) Use the equation $v = u + at$, with $u = 30$, $v = 0$ and $a = -5$ to find t.

$$0 = 30 + (-5)t$$

$$t = \frac{30}{5} = 6$$

Using **6** (i)

(c) Use the equation $v^2 = u^2 + 2as$, with $u = 30$, $v = 0$ and $a = -5$ to find s.

$$0^2 = 30^2 + 2 \times (-5)s$$

$$0 = 900 - 10s$$

$$s = 90 \text{ m}$$

Using **6** (iv)

Worked example 3

A ball is thrown vertically upwards with an initial speed of 7 m s^{-1} from a height of 2 metres.

(a) Find the maximum height of the ball above ground level.

(b) Find the speed of the ball when it hits the ground.

(c) Find the time that it takes the ball to reach the ground.

(a) When the ball reaches its maximum height $v = 0$. Using the formula $v^2 = u^2 + 2as$, with $u = 7$ and $a = -9.8$ gives,

$$0^2 = 7^2 + 2 \times (-9.8) \times s$$

$$0 = 49 - 19.6s$$

$$s = \frac{49}{19.6} = 2.5 \text{ m}$$

As the ball is initially 2 metres above ground level,

maximum height $= 2.5 + 2 = 4.5 \text{ m}$

Using **6** (iv)

(b) When the ball hits the ground $s = -2$, as it will be 2 metres below its initial position. Using the equation $v^2 = u^2 + 2as$ to find v gives,

$$v^2 = 7^2 + 2 \times (-9.8) \times (-2)$$
$$= 88.2$$
$$v = \sqrt{88.2} = 9.39 \text{ m s}^{-1} \text{ (to 3 s.f.)}$$

Using **6** (iv)

Note that the positive value is taken as the question asks for the speed.

(c) Using the formula $s = \frac{1}{2}(u + v)t$, with $s = -2$, $u = 7$ and $v = -\sqrt{88.2}$ gives,

$$-2 = \frac{1}{2}(7 - \sqrt{88.2})t$$

$$t = \frac{-2 \times 2}{7 - \sqrt{88.2}} = 1.67 \text{ s (to 3 s.f.)}$$

Using **6** (iii)

Note that the ball will be moving downwards when it hits the ground, so that $v = -\sqrt{88.2}$.

REVISION EXERCISE I

1 The graph shows how the velocity of a train varies over a period of 200 seconds along a length of straight horizontal track.

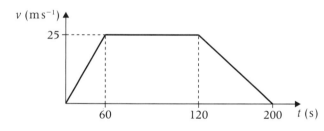

(a) Find the total distance travelled by the train.
(b) Find the average speed of the train.
(c) Find the acceleration of the train when $120 < t < 200$.

2 The graph shows how the velocity of a cyclist varies as she moves along a straight line for a period of 40 seconds.

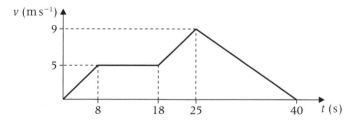

(a) Find the total distance travelled by the cyclist.
(b) Calculate the average speed of the cyclist.
(c) Find the maximum acceleration of the cyclist.

3 A shunting engine moves backwards and forwards on a straight track. The graph below shows how the velocity of the engine varies during a 100 second period of time.

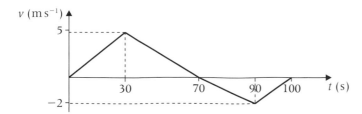

(a) Find the total distance travelled by the engine.

(b) Calculate the average speed of the engine.

(c) Find the distance of the engine from its starting point at the end of the 100 second period.

(d) Calculate the average velocity of the engine.

4 A particle moves on a straight line, so that its velocity is as shown in the graph below.

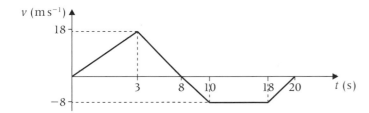

(a) Find the total distance travelled by the particle.

(b) Calculate the average speed of the particle.

(c) Find the distance of the particle from its initial position at the end of the 20 second period.

(d) Calculate the average velocity of the particle during the 20 second period.

5 The graph below shows a velocity–time graph for a particle that is moving on a straight line.

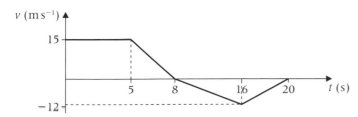

(a) Find the total distance travelled by the particle.

(b) Find the average speed of the particle.

(c) Find the average velocity of the particle.

6 The points A and B are on a straight line. Henry walks from A to B in 5 seconds travelling at a constant speed of $4 \, \text{m s}^{-1}$. Ben walks from A to B starting at rest, but moving with a constant acceleration. Ben also takes 5 seconds to travel from A to B.

  (a) Find the distance between A and B.

  (b) Find the speed at which Ben is moving when he arrives at B.

  (c) Find Ben's acceleration.

7 A car travels a distance of 100 metres along a straight line in 20 seconds. The car was initially at rest and travelled with uniform acceleration.

  (a) Find the acceleration of the car.

  (b) Find the speed of the car when it has travelled 100 metres.

8 The speed of a cyclist moving on a straight line, with a constant acceleration, increases from $4 \, \text{m s}^{-1}$ to $8 \, \text{m s}^{-1}$ in a 20 second period of time.

  (a) Find the acceleration of the cyclist.

  (b) Find the distance that she travels in the 20 second period.

9 A particle moves along a straight line with a constant acceleration. Initially the particle has a velocity of $16 \, \text{m s}^{-1}$ and is at the origin. After it has been accelerating for 20 seconds its velocity is $-10 \, \text{m s}^{-1}$.

  (a) Find the acceleration of the particle.

  (b) Find the distance of the particle from the origin when its velocity is zero.

  (c) Find the time that it takes the particle to move from the origin to the point where its velocity is zero.

  (d) Find the distance of the particle from the origin when its velocity is $-10 \, \text{ms}^{-1}$.

10 As a lift moves between two floors, its motion has three stages. In stage one the lift accelerates from rest at $0.4 \, \text{m s}^{-2}$ until it reaches a speed of $1.2 \, \text{m s}^{-1}$. In stage 2 the lift travels with a constant velocity for 12 seconds. In stage three the lift decelerates uniformly for 5 seconds before coming to rest.

  (a) Find the total time for which the lift is moving.

  (b) Find the total distance travelled by the lift.

  (c) Find the average speed of the lift.

11 A car is travelling on a straight road. It is travelling at 30 m s^{-1} when it begins to brake. The car decelerates uniformly and comes to rest when it has travelled 90 metres.

(a) Find the acceleration of the car.

(b) Find the time that the car takes to stop.

(c) Find the speed of the car when it has been braking for 5 seconds.

(d) Find the distance the car has moved as its speed is reduced from 30 m s^{-1} to 20 m s^{-1}.

12 A cyclist applies her brakes when travelling at 5 m s^{-1}. She decelerates uniformly stopping after she has travelled 10 metres along a straight line. Find the time it takes for the cyclist to reduce her speed to 2 m s^{-1} and the distance that she travels as this happens.

13 A ball is projected vertically upwards from ground level at a speed of 28 m s^{-1}.

(a) Find the maximum height of the ball.

(b) Find the time that the ball is in the air.

14 A ball is thrown vertically upwards with a speed of 5 m s^{-1}, from a point at a height of 8 metres above ground level.

(a) Find the time it takes the ball to reach its maximum height.

(b) Find the maximum height of the ball above ground level.

(c) Find the time that it takes the ball to reach ground level.

(d) Find the speed of the ball when it hits the ground.

15 A ball is released from rest at a point at a height of 6 metres above ground level.

(a) Find the time that it takes the ball to reach ground level.

(b) Find the speed of the ball when it hits the ground.

16 A ball is projected vertically upwards from ground level. It returns to ground level 1.8 seconds later.

(a) Find the initial speed of the ball.

(b) Find the maximum height of the ball above ground level.

17 A stone is allowed to fall from rest from a point at a height h metres above ground level. The stone hits the ground travelling at 12 m s^{-1}.

(a) Find h.

(b) Find the time that the ball takes to fall to ground level.

18 A particle is projected vertically upwards from ground level with a speed of 21 m s^{-1}.

   (a) Find the maximum height of the particle.

   (b) Find the time for which the particle is in the air.

   (c) Find the length of time for which the height of the particle above ground level is greater than 5 metres.

19 A car accelerates from rest at 1.5 m s^{-2} for T seconds and then travels at a constant speed. The car travels a total of 312 metres, on a straight line, in 30 seconds.

   (a) Find T.

   (b) Find the constant speed at which the car travels after it has stopped accelerating.

20 A particle moves on a straight line. Initially it has a velocity of 20 m s^{-1} and is at the origin. The particle experiences an acceleration of -0.4 m s^{-2} for 30 seconds and then moves with a constant velocity for a further 60 seconds.

   (a) Find the velocity of the particle 30 seconds after it left the origin.

   (b) Find the distance of the particle from the origin at the end of the 90 second period.

21 A ball is projected vertically upwards with speed U m s^{-1}. It reaches a maximum height of 3 metres above the point of projection.

   (a) Find U.

   (b) Find the time that it takes the ball to reach its maximum height.

22 A bullet is fired vertically upwards with an initial speed of 120 m s^{-1} from a height of 2 metres above ground level. The bullet hits the roof of a barn, at a height of 12 metres.

   (a) Find the speed of the bullet when it hits the roof.

   (b) Find the time that it takes for the bullet to reach the roof.

23 On the moon the acceleration due to gravity is 1.6 m s^{-2}. A ball is released from a height of 3 metres.

   (a) Find the time that it takes the ball to reach the ground.

   (b) Find the speed of the ball when it hits the ground.

24 On a planet a ball is allowed to fall from rest. It takes 2 seconds for a ball to fall 4 metres to ground level.

  (a) Find the magnitude of the acceleration due to gravity on this planet.

  (b) Find the speed of the ball when it hits the ground.

25 A particle moves on a straight line with uniform acceleration of -1.2 m s^{-2}. The initial velocity of the particle is 6 m s^{-1}. Find the three times when the particle is at a distance of 2 metres from its initial position.

26 A car moves on a straight line with constant acceleration, so that at time t seconds its velocity is $v \text{ m s}^{-1}$. When $t = 10$, $v = 12$ and when $t = 14$, $v = 19$. At time $t = 0$, the particle is at the origin.

  (a) Find the acceleration of the particle.

  (b) Find the velocity of the particle at time t.

  (c) Find the distance of the particle from the origin at time t.

  (d) Find the times at which the particle is at the origin.

27 A train accelerates uniformly from rest along a straight horizontal track. After it has travelled 400 metres, it speed is 16 m s^{-1}.

  (a) **(i)** Show that the acceleration of the train is 0.32 m s^{-2}.
      (ii) Find the time that it takes the train to travel the 400 metres.

  (b) When the train has reached a speed of 16 m s^{-1}, its acceleration is increased to 0.5 m s^{-2}.
      (i) Find the distance that the train travels as its speed increases from 16 m s^{-1} to 30 m s^{-1}.
      (ii) Find the total time that the train has been moving when it reaches a speed of 30 m s^{-1}.   [A]

28 The velocity–time graph below models Tom's motion as he runs across a playground.

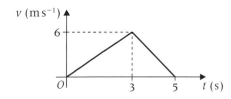

  (a) Find Tom's acceleration during the first 3 seconds of his motion.

  (b) Find the total distance Tom runs during the 5 seconds of motion.

  (c) State one criticism of this model of Tom's motion.   [A]

29 A particle *P* moves in a straight line across a horizontal surface with retardation of magnitude 1.8 m s⁻². The particle *P* passes through a point *A* with velocity 5 m s⁻¹, as shown in the diagram. It subsequently passes twice through the point *B*, where *AB* = 2.5 metres.

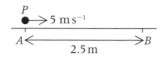

(a) Find the velocities of *P* on the two occasions when it passes through *B*.

(b) Hence, or otherwise, find the length of time between the two occasions when *P* passes through *B*.   [A]

Test yourself	**What to review**
	If your answer is incorrect:
1 A car moves with a constant acceleration of 2 m s⁻² along a straight line. The car starts from rest.	See p 2 Example 2 or review Advancing Maths for AQA M1 pp 18–19
(a) Find the time that it would take the car to reach a speed of 14 m s⁻¹.	
(b) Find the distance that the car would have travelled by the time it reached this speed.	
2 A train accelerates uniformly from 12 m s⁻¹ to 30 m s⁻¹ in 3 minutes as it travels along a straight line.	See p 1 Example 1 or review Advancing Maths for AQA M1 pp 20–21
(a) Find the acceleration of the train.	
(b) Find the distance travelled by the train as it accelerates.	
(c) Find the time it takes for the train to increase in speed from 12 m s⁻¹ to 20 m s⁻¹.	
3 A ball is released from rest at a height of 10 metres.	See p 3 Example 3 or review Advancing Maths for AQA M1 pp 25–26
(a) Find the speed of the ball when it hits the ground.	
(b) Find the time that it takes the ball to reach the ground.	
4 A lorry is travelling at *U* m s⁻¹ when it begins to accelerate. In the first 4 seconds that it is accelerating it travels 24 metres and when it stops accelerating after 10 seconds it has travelled 105 m.	See p 1 Example 1 or review Advancing Maths for AQA M1 pp 18–21
(a) Find *U*.	
(b) Find the acceleration of the lorry.	

5 A small stone falls vertically from rest. When the stone hits the ground it is travelling at a speed of 24.5 m s^{-1}. Model the stone as a particle and assume that no resistance forces act on the stone as it falls.

   See p 3 Example 3 or review Advancing Maths for AQA M1 pp 25–26

(a) Find the time that it takes for the stone to fall to the ground.

(b) Show that the stone falls a distance of 30.625 metres.

(c) Find the time for which the stone has been falling when it is 5 metres above ground level. [A]

CHAPTER 2

Kinematics in two dimensions

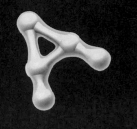

Key points to remember

1 $\mathbf{v} = \mathbf{u} + \mathbf{a}t$

2 $\mathbf{r} = \mathbf{u}t + \dfrac{1}{2}\mathbf{a}t^2$ or $\mathbf{r} = \mathbf{u}t + \dfrac{1}{2}\mathbf{a}t^2 + \mathbf{r}_0$

3 $\mathbf{r} = \dfrac{1}{2}(\mathbf{u} + \mathbf{v})t$ or $\mathbf{r} = \dfrac{1}{2}(\mathbf{u} + \mathbf{v})t + \mathbf{r}_0$

Where:
$\mathbf{a}$ = acceleration
$\mathbf{v}$ = velocity at time t
$\mathbf{u}$ = initial velocity
$\mathbf{r}$ = position vector at time t
$\mathbf{r}_0$ = initial position vector

Worked example 1

A ship is travelling with a constant velocity of $(4\mathbf{i} + 5\mathbf{j})$ m s^{-1} when it begins to accelerate. It experiences an acceleration of $(0.2\mathbf{i} - 0.5\mathbf{j})$ m s^{-1} for 20 seconds.

(a)   Find the velocity and speed of the ship at the end of the 20 seconds.

(b)   Assume that the ship is at the origin when it begins to accelerate and find its position vector at the end of the 20 seconds.

(a)   Use $\mathbf{v} = \mathbf{u} + \mathbf{a}t$ to find the velocity at the end of the 20 seconds.

$$\mathbf{v} = (4\mathbf{i} + 5\mathbf{j}) + (0.2\mathbf{i} - 0.5\mathbf{j}) \times 20$$
$$= 4\mathbf{i} + 5\mathbf{j} + 4\mathbf{i} - 10\mathbf{j}$$
$$= 8\mathbf{i} - 5\mathbf{j}$$

The speed will be the magnitude of the velocity.

$$v = \sqrt{8^2 + 5^2} = 9.43 \text{ m s}^{-1} \text{ (to 3 s.f.)}$$

Using **1**

(b)   Use $\mathbf{r} = \mathbf{u}t + \dfrac{1}{2}\mathbf{a}t^2$ to find the position vector when $t = 20$.

$$\mathbf{r} = (4\mathbf{i} + 5\mathbf{j}) \times 20 + \frac{1}{2}(0.2\mathbf{i} - 0.5\mathbf{j}) \times 20^2$$
$$= 80\mathbf{i} + 100\mathbf{j} + 40\mathbf{i} - 100\mathbf{j}$$
$$= 120\mathbf{i}$$

Using **2**

Worked example 2

A particle moves with constant velocity, so that during a 10 second period of time its velocity changes from $(9\mathbf{i} - 2\mathbf{j})$ m s^{-1} to $(-\mathbf{i} + 23\mathbf{j})$ m s^{-1}. The unit vectors $\mathbf{i}$ and $\mathbf{j}$ are directed east and north respectively.

(a) Find the acceleration of the particle.

(b) Find the speed of the particle when it is moving due east.

(a) Substitute into the formula $\mathbf{v} = \mathbf{u} + \mathbf{a}t$ and solve to find the acceleration.

$$-\mathbf{i} + 23\mathbf{j} = 9\mathbf{i} - 2\mathbf{j} + 10\mathbf{a}$$
$$10\mathbf{a} = -10\mathbf{i} + 25\mathbf{j}$$
$$\mathbf{a} = -\mathbf{i} + 2.5\mathbf{j}$$

Using **1**

(b) Taking $t = 0$ when the velocity is $(9\mathbf{i} - 2\mathbf{j})$ m s^{-1}, gives the expression below for the velocity of the particle.

$$\mathbf{v} = (9\mathbf{i} - 2\mathbf{j}) + (-\mathbf{i} + 2.5\mathbf{j})t$$
$$= (9 - t)\mathbf{i} + (-2 + 2.5t)\mathbf{j}$$

Using **1**

When the particle is travelling due east, the $\mathbf{j}$ component of the velocity will be zero.

$$-2 + 2.5t = 0$$
$$t = \frac{2}{2.5} = 0.8$$

The velocity at this time can now be found.

$$\mathbf{v}(0.8) = (9 - 0.8)\mathbf{i} = 8.2\mathbf{i}$$

The speed of the particle will be 8.2 m s^{-1}.

Worked example 3

A particle moves with a constant acceleration of $(\mathbf{i} + 2\mathbf{j})$ m s^{-2}. At time $t = 0$, the position vector of the particle is $(40\mathbf{i} - 4\mathbf{j})$ metres and its velocity is $(10\mathbf{i} + \mathbf{j})$ m s^{-1}. The unit vectors are directed east and north respectively.

(a) Find an expression for the position vector of the particle at time t seconds.

(b) Find the time when the particle is north east of the origin.

(c) Find the distance of the particle from the origin at this time.

(a) Use the constant acceleration formula $\mathbf{r} = \mathbf{u}t + \frac{1}{2}\mathbf{a}t^2 + \mathbf{r}_0$ to find the position vector.

Using **2**

$$\mathbf{r} = (10\mathbf{i} + \mathbf{j})t + \frac{1}{2}(\mathbf{i} + 2\mathbf{j})t^2 + (40\mathbf{i} - 4\mathbf{j})$$

$$= 10t\mathbf{i} + t\mathbf{j} + \frac{1}{2}t^2\mathbf{i} + t^2\mathbf{j} + 40\mathbf{i} - 4\mathbf{j}$$

$$= \left(10t + \frac{1}{2}t^2 + 40\right)\mathbf{i} + (t + t^2 - 4)\mathbf{j}$$

(b) When the particle is north east of the origin the $\mathbf{i}$ and $\mathbf{j}$ components of the position vector will be equal. This gives an equation which can then be solved.

Using **2**

$$10t + \frac{1}{2}t^2 + 40 = t + t^2 - 4$$

$$\frac{1}{2}t^2 - 9t - 44 = 0$$

$$t^2 - 18t - 88 = 0$$

$$(t + 4)(t - 22) = 0$$

$$t = -4 \text{ or } t = 22$$

As we require a positive time we take $t = 22$.

Substituting $t = 22$ gives,

$$\mathbf{r} = \left(10 \times 22 + \frac{1}{2} \times 22^2 + 40\right)\mathbf{i} + (22 + 22^2 - 4)\mathbf{j}$$

$$= 502\mathbf{i} + 502\mathbf{j}$$

This confirms that the particle is in fact north east of the origin and not south west, which would have been possible if the components had been negative.

(c) The distance from the origin is simply given by the magnitude of the position vector.

$$\text{Distance} = \sqrt{502^2 + 502^2} = 710 \text{ m (to 3 s.f.)}$$

Worked example 4

An aeroplane flies due north at a speed of 140 m s^{-1} relative to the air. Due to a wind the air is moving due west at a speed of 60 m s^{-1}. Find the resultant velocity of the aeroplane.

The diagram shows the velocities involved in this problem.

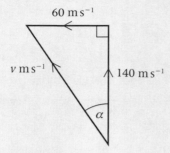

Pythagoras' Theorem can be used to find v, the magnitude of the resultant velocity.

$$v^2 = 140^2 + 60^2$$
$$= 23\,200$$
$$v = \sqrt{23\,200} = 152 \text{ m s}^{-1} \text{ (to 3 s.f.)}$$

The angle α can be found using trigonometry.

$$\tan \alpha = \frac{60}{140}$$
$$\alpha = \tan^{-1}\left(\frac{60}{140}\right) = 23.2°$$

This direction can be given as a bearing $360 - 23.2 = 336.8°$.

Worked example 5

A canoe is paddled in a river that is flowing at 4 m s^{-1}. The velocity of the canoe relative to the water is 5 m s^{-1} at an angle of $60°$ to the upstream bank of the river.

(a) Find the magnitude of the resultant velocity of the canoe.

(b) Find the direction of the resultant velocity, giving your answer as the angle to the upstream bank.

The diagram shows the velocities of the water, the canoe relative to the water and the resultant velocity of the canoe, which has magnitude v and is at an angle of α to the downstream bank.

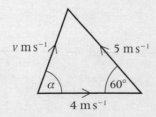

(a) The cosine rule can be used in this triangle to find v.

$$v^2 = 5^2 + 4^2 - 2 \times 4 \times 5 \cos 60°$$
$$= 21$$
$$v = \sqrt{21} = 4.58 \text{ m s}^{-1} \text{ (to 3 s.f.)}$$

(b) The sine rule can be used to find α.

$$\frac{\sin \alpha}{5} = \frac{\sin 60°}{\sqrt{21}}$$

$$\sin \alpha = \frac{5 \sin 60°}{\sqrt{21}}$$

$$\alpha = \sin^{-1}\left(\frac{5 \sin 60°}{\sqrt{21}}\right) = 70.9°$$

The angle to the upstream bank will be $180 - 70.9 = 109.1°$.

REVISION EXERCISE 2

1 A particle moves with a constant acceleration of $(4\mathbf{i} - 2\mathbf{j})$ m s^{-2}. At time t seconds the velocity of the particle is $\mathbf{v}$ and its position vector is $\mathbf{r}$. When $t = 0$, $\mathbf{v} = (5\mathbf{i} + 6\mathbf{j})$ m s^{-1} and the particle is at the origin.

  (a) Find $\mathbf{r}$ when $t = 10$.

  (b) Hence find the distance of the particle from the origin when $t = 10$.

  (c) Find $\mathbf{v}$ when $t = 10$.

  (d) Hence find the speed of the particle when $t = 10$.

2 At time $t = 0$, a boat is heading due east at 4 m s^{-1} and is at the origin. The boat moves with an acceleration of $(\mathbf{i} + \mathbf{j})$ m s^{-2}. The unit vectors $\mathbf{i}$ and $\mathbf{j}$ are directed east and north respectively.

  (a) Write down the initial velocity of the boat as a vector.

  (b) Find the velocity of boat when $t = 6$ seconds.

  (c) Find the position vector of the boat when $t = 5$ seconds.

3 The velocity, $\mathbf{v}$ m s^{-1}, of a particle at time t seconds is given by

$$\mathbf{v} = (8 - 6t)\mathbf{i} + (2 + 4t)\mathbf{j}.$$

Find the speed of the particle when $t = 10$ seconds.

4 The position vector, $\mathbf{r}$ metres, of a jet ski at time t seconds is defined below.

$$\mathbf{r} = (4 + t + t^2)\mathbf{i} + (2t^2)\mathbf{j}$$

(a) Find the position vector of the jet ski when $t = 0$.

(b) Find the position vector of the jet ski when $t = 5$.

(c) Find the distance between the two positions that you have calculated.

5 At time $t = 0$ a particle has velocity $(\mathbf{i} - 2\mathbf{j})$ m s^{-1} and position vector $(30\mathbf{i} + 18\mathbf{j})$ metres. It moves with a constant acceleration of $(2\mathbf{i} + \mathbf{j})$ m s^{-2}.

(a) Find an expression for the velocity of the particle at time t seconds.

(b) Find t when the velocity is $(7\mathbf{i} + \mathbf{j})$ m s^{-1}.

(c) Find an expression for the position vector of the particle at time t seconds.

(d) Find t when the position vector is $(140\mathbf{i} + 48\mathbf{j})$ metres.

6 A boat has initial velocity $(8\mathbf{j})$ m s^{-1} and an acceleration of $(2\mathbf{i})$ m s^{-2}. The unit vectors $\mathbf{i}$ and $\mathbf{j}$ are directed east and north respectively.

(a) Find an expression for the velocity of the boat at time t seconds.

(b) Find the time when the boat is travelling north east.

7 A particle has initial velocity $(3\mathbf{i} + 8\mathbf{j})$ m s^{-1} and 20 seconds later has velocity $(43\mathbf{i} + 70\mathbf{j})$ m s^{-1}.

(a) Find the acceleration of the particle.

(b) Initially the particle is at the origin. Find the distance of the particle from the origin 20 seconds later.

8 In a 10 second period the velocity of a helicopter changes uniformly from $(10\mathbf{i} + 8\mathbf{j})$ m s^{-1} at time $t = 0$ to $(6\mathbf{i} - 32\mathbf{j})$ m s^{-1} at time $t = 10$ seconds. The unit vectors $\mathbf{i}$ and $\mathbf{j}$ are directed east and north respectively.

(a) Find the acceleration of the helicopter.

(b) Find an expression for the velocity of the helicopter at time t seconds.

(c) Find the speed of the helicopter when it is flying due east.

9 A jet ski moves from rest at the origin to the point with position vector $(40\mathbf{i} + 160\mathbf{j})$ metres in 20 seconds. The jet ski moves with a constant acceleration.

(a) Find the distance of the jet ski from its initial position.

(b) Find the acceleration of the jet ski.

(c) Find the velocity and speed of the jet ski when it has been moving for 20 seconds.

10 A lighthouse is taken as the origin to find the position vector of a boat. The unit vectors $\mathbf{i}$ and $\mathbf{j}$ are directed east and north respectively. At time t seconds the velocity of the boat is $\mathbf{v}$ and the position vector is $\mathbf{r}$. At time $t = 0$, $\mathbf{v} = (3\mathbf{i} + 2\mathbf{j})$ m s^{-1} and $\mathbf{r} = (-80\mathbf{i} - 10\mathbf{j})$ metres. The boat has a constant acceleration of $(0.1\mathbf{i} - 0.2\mathbf{j})$ m s^{-2}.

  (a) Show that $\mathbf{r} = (0.05t^2 + 3t - 80)\mathbf{i} + (-0.1t^2 + 2t - 10)\mathbf{j}$.

  (b) Show that the boat is due south of the lighthouse when $t = 20$.

  (c) Find the time when the boat is due west of the lighthouse.

11 A particle is initially at rest at the point with position vector $(9\mathbf{i} + 5\mathbf{j})$ m. At time t seconds the velocity of the particle is $\mathbf{v}$ and the position vector is $\mathbf{r}$. The particle experiences a constant acceleration of $(0.3\mathbf{i} - 0.4\mathbf{j})$ m s^{-2}. The unit vectors $\mathbf{i}$ and $\mathbf{j}$ are directed east and north respectively.

  (a) Express $\mathbf{v}$ in terms of t.

  (b) Find the value of t for which the speed of the particle is 9 m s^{-1}.

  (c) Express $\mathbf{r}$ in terms of t.

  (d) Hence find the time when the particle is due east of the origin.

12 A particle has velocity $(2\mathbf{i} + 5\mathbf{j})$ m s^{-1} at time $t = 0$ and moves with acceleration $(-0.1\mathbf{i} - 0.4\mathbf{j})$ m s^{-2}.

  (a) Find an expression for the velocity of the particle at time t.

  (b) Find the time when the particle is travelling due east.

  (c) Find the time when the particle is travelling due south.

13 A particle moves, so that at time t seconds its velocity is $\mathbf{v}$ m s^{-1}. When $t = 3$, $\mathbf{v} = (4\mathbf{i} + 9\mathbf{j})$ m s^{-1} and when $t = 8$, $\mathbf{v} = (24\mathbf{i} - 11\mathbf{j})$ m s^{-1}. The unit vectors $\mathbf{i}$ and $\mathbf{j}$ are directed east and north respectively.

  (a) Find the acceleration of the particle.

  (b) Find the velocity of the particle when $t = 0$.

  (c) Find an expression for the velocity of the particle at time t seconds.

  (d) Find the time when the particle is travelling due east.

14 A particle moves in a horizontal plane, in which the unit vectors $\mathbf{i}$ and $\mathbf{j}$ lie. These unit vectors are directed east and north respectively. Initially the particle is at the origin. It is 40 metres due south of the origin 4 seconds after it leaves the origin and 10 seconds after it has left the origin its position vector is $(-60\mathbf{i} - 280\mathbf{j})$ metres.

(a) Find the acceleration of the particle.

(b) Find the velocity of the particle when it is at the origin.

(c) Find the time when the particle is travelling due south.

15 A particle is initially at rest at the origin. After accelerating for 6 seconds its velocity is $(3.6\mathbf{i} - 4.8\mathbf{j})$ m s^{-1}.

(a) Find the acceleration of the particle.

(b) Find the distance of the particle from the origin after it has been accelerating for 6 seconds.

16 A boat moves from the origin where it was at rest at time $t = 0$ and moves with a constant acceleration $\mathbf{a}$ m s^{-2}. When $t = 12$ seconds the position vector of the boat is $(2.88\mathbf{i} + 4.32\mathbf{j})$.

(a) Find $\mathbf{a}$.

(b) The position vector of the boat is $(32\mathbf{i} + 48\mathbf{j})$ metres, T seconds after leaving the origin. Find T.

17 The position vector of a particle at time t is $\mathbf{r}$, where

$$\mathbf{r} = (5t - 2t^2)\mathbf{i} + (8t - 3t^2)\mathbf{j}$$

(a) Show that the particle was initially at the origin.

(b) By writing $\mathbf{r}$ in the form $\mathbf{r} = \mathbf{u}t + \frac{1}{2}\mathbf{a}t^2$, find $\mathbf{u}$ and $\mathbf{a}$.

18 The width of a river is 20 metres. The current in the river flows at 2 m s^{-1}. A boat which moves at a speed of 4 m s^{-1} relative to the water, sets off from one bank. The resultant velocity of the boat is at right angles to the bank.

(a) Find the magnitude of the resultant velocity of the boat.

(b) Find the time that it takes for the boat to cross the river.

19 The velocity of an aeroplane, relative to the air, is directed due north and has magnitude 80 m s^{-1}. A wind is blowing due east with a constant speed of 12 m s^{-1}.

(a) Find the magnitude of the resultant velocity of the aeroplane.

(b) Find the direction of the resultant velocity of the aeroplane, giving your answer as a bearing correct to the nearest degree.

20 An aeroplane travels due north at 120 m s^{-1}, in air which is moving due west at a speed of 20 m s^{-1}. Find the magnitude and direction of the velocity of the aeroplane relative to the air.

21 The resultant velocity of a ship has magnitude 8 m s^{-1} and is directed due south. A current is flowing south west with speed 2 m s^{-1}.

(a) Find the magnitude of the velocity of the ship relative to the water.

(b) Find the direction of this relative velocity, giving your answer as a bearing.

22 A boat travels due east at 6 m s^{-1} relative to the water in which it is moving. A current flows south west at 3 m s^{-1}. Find the magnitude and direction of the resultant velocity of the boat.

23 A child swims across a river. Her velocity relative to the water is at right angles to the bank and has magnitude 2.4 m s^{-1}. The water in the river flows at 1.2 m s^{-1}.

(a) Find the magnitude of the resultant velocity of the child.

(b) Find the direction of the resultant velocity of the child, giving your answer as the angle that it makes to the downstream bank.

24 A canoe is paddled across a river, in which the water flows at 1 m s^{-1}. The canoe has a velocity relative to the water of 3 m s^{-1} at an angle α to the upstream bank of the river. The resultant velocity is at an angle of $45°$ to the upstream bank.

(a) Find α.

(b) Find the magnitude of the resultant velocity of the canoe.

25 The diagram shows the resultant velocity of a boat that is crossing a river and the current in the river.

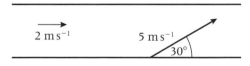

Find the magnitude and direction of the velocity of the boat relative to the water.

26 The resultant velocity of an aeroplane is 120 m s^{-1} on a bearing of $120°$. It is flying in air that is moving at 35 m s^{-1} on a bearing of $230°$. Find the magnitude and direction of the velocity of the aeroplane relative to the air.

27 The unit vectors **i** and **j** are directed east and north respectively. A yacht moves with a constant acceleration. At time t seconds the position vector of the yacht is **r** metres.

When $t = 0$ the velocity of the yacht is $(2\mathbf{i} - \mathbf{j})\,\text{m s}^{-1}$ and when $t = 10$ the velocity of the yacht is $(-\mathbf{i} + \mathbf{j})\,\text{m s}^{-1}$.

(a) Find the acceleration of the yacht.

(b) When $t = 0$ the yacht is 20 metres due east of the origin. Find **r**.

(c) (i) Show that when $t = 20$ the yacht is due north of the origin.

   (ii) Find the speed of the yacht when t = 20.   [A]

28 A model aeroplane moves in a horizontal plane with a constant acceleration. Initially the aeroplane is at the origin and has velocity $(35\mathbf{i} + 45\mathbf{j})\,\text{m s}^{-1}$. After accelerating for 8 seconds, the velocity of the aeroplane is $(19\mathbf{i} + 13\mathbf{j})\,\text{m s}^{-1}$. The unit vectors **i** and **j** are perpendicular and lie in the horizontal plane.

(a) Show that the acceleration of the aeroplane is $(-2\mathbf{i} - 4\mathbf{j})\,\text{m s}^{-2}$.

(b) Find an expression for the position vector of the aeroplane at time t seconds.

(c) Find the time when the position vector of the aeroplane is $(300\mathbf{i} + 225\mathbf{j})\,\text{m}$.   [A]

29 A motor boat can travel at a speed of $6\,\text{m s}^{-1}$ relative to the water. It is used to cross a river in which the current flows at $2\,\text{m s}^{-1}$. The resultant velocity of the boat makes an angle of $60°$ to the river bank as shown in the diagram below.

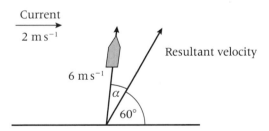

The angle between the direction in which the boat is travelling and the resultant velocity is α.

(a) Show that $\alpha = 16.8°$, correct to three significant figures.

(b) Find the magnitude of the resultant velocity.   [A]

30 Jane is playing with her toy motor boat. She releases the boat from the point C on one bank of a stream. The boat travels to the point D on the other bank of the stream.

Until it reaches D, the boat has constant velocity $2\mathbf{i} - 6\mathbf{j}\,\text{m s}^{-1}$ relative to the stream. The stream moves with constant velocity $3\mathbf{i}\,\text{m s}^{-1}$, affecting the motion of the boat.

The banks of the stream are parallel to **i**, as shown in the diagram below.

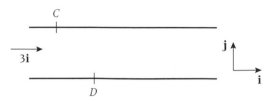

(a) Find the resultant velocity of the boat as it travels from C to D.

(b) Jane's sister turns the boat around and releases it from D so that it travels directly back to Jane at C. The boat travels along the straight line DC at the same speed as on the journey from C to D. The stream still moves with constant velocity $3\mathbf{i}$ m s^{-1}.

   (i) State the new resultant velocity of the boat as it travels back across the stream.

  (ii) Find the magnitude of the constant velocity, relative to the stream, with which the boat travels from D to C.   [A]

Test yourself	What to review
	If your answer is incorrect:
1 At time $t = 0$ the velocity of a skater is $(0.3\mathbf{i} + 0.7\mathbf{j})$ m s^{-1}. She accelerates uniformly, so that when $t = 4$ seconds her velocity is $(\mathbf{i} + 1.9\mathbf{j})$ m s^{-1}. (a) Find an expression for her velocity at time t. (b) Taking her position at time $t = 0$ as the origin, find an expression for her position at time t.	See pp 12–13 Examples 1 and 3 or review Advancing Maths for AQA M1 pp 42–44
2 A ship moves so that its position vector, $\mathbf{r}$, at time t is given by $\mathbf{r} = (9 - 2t)\mathbf{i} + (24 - 3t)\mathbf{j}$ The unit vectors $\mathbf{i}$ and $\mathbf{j}$ are directed east and north respectively. (a) Find t when the ship is due north of the origin. (b) Find t when the ship is due west of the origin.	See p 13 Example 3 or review Advancing Maths for AQA M1 pp 42–44
3 A particle moves with constant acceleration between the two points, A and B, with position vectors $\mathbf{r}_A = (5\mathbf{i} + 9\mathbf{j})$ metres and $\mathbf{r}_B = (-5\mathbf{i} + 39\mathbf{j})$ in a 5 second period. At time $t = 0$ the particle is at rest at A. Find the velocity of the particle when it reaches B.	See p 13 Example 2 or review Advancing Maths for AQA M1 pp 42–44
4 A ship intends to travel due east. A current flows due south at 0.5 m s^{-1}. The speed of the ship is 8 m s^{-1} on a bearing of $\alpha°$ relative to the water in which it is moving. The resultant velocity of the ship is due east.	See pp 14–15 Examples 4 and 5 or review Advancing Maths for AQA M1 pp 48–50

(a) Find α.
(b) Find the magnitude of the resultant velocity of the ship.

---

5 A boat sails on a bearing of $\alpha°$ with a constant speed of x m s^{-1} relative to the water. A tidal current acts at a constant speed of 1.2 m s^{-1} on a bearing of 090°.

The resultant speed of the boat is y m s^{-1} on a bearing of 030° along the line *AB*, as shown in the diagram.

See pp 14–15 Examples 4 and 5 or review Advancing Maths AQA M1 pp 48–50

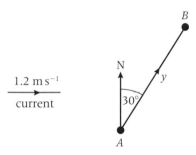

(a) The distance *AB* is 120 metres and the boat takes 20 seconds to sail from *A* to *B*. Find the value of *y*.
(b) Draw an appropriate triangle of velocities.
(c) Find the northerly and easterly components of *y*.
(d) Hence or otherwise:
    (i) show that the value of *x* is approximately 5.50 m s^{-1};
    (ii) find the value of α.   [A]

---

5 (a) $y = 6$
(c) 5.20 and 3
(d) (ii) 19.1°

4 (a) 86.4°
(b) 7.98 m s^{-1}

3 $(-4\mathbf{i} + 12\mathbf{j})$ m s^{-1}

2 (a) $t = 4.5$
(b) $t = 8$

1 (a) $(0.7 + 0.175t)\mathbf{i} + (0.3 + 0.3t)\mathbf{j}$
(b) $(0.7t + 0.0875t^2)\mathbf{i} + (0.3t + 0.15t^2)\mathbf{j}$

CHAPTER 3

Forces

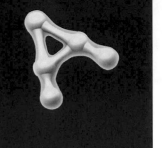

Key points to remember

1 The weight is given by mg where m is the mass of the body and g is that acceleration due to gravity, which is 9.8 m s^{-2} on Earth.

2 $F \leqslant \mu R$, where F is the magnitude of the friction force, R is the magnitude of the normal reaction force and μ is the coefficient of friction.

3 When an object is in contact with a surface there will be a normal reaction force at right angles to the surface and a friction force parallel to the surface.

Note that the diagram also shows the weight acting on the body.

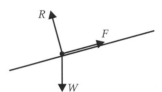

4 If the forces acting on a body are in equilibrium, their resultant will be zero.

5 The resultant force is the sum of all of the forces that are acting.

6 If a body remains at rest or moves with constant velocity the forces acting on it will be in equilibrium.

7 The diagram shows the two components of a force.

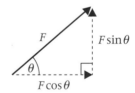

Worked example 1

The diagram shows the forces acting on a particle of mass 5 kg, which is at rest on a slope inclined at 5° to the horizontal.

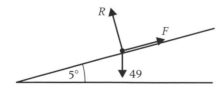

(a) Find R.

(b) Find F.

(c) Find an inequality that must be satisfied by μ, the coefficient of friction between the particle and the slope.

(a) Resolving perpendicular to the slope gives,

$$R = 49 \cos 5° = 48.8 \text{ N} \text{ (to 3 s.f.)}$$

Using **6**

(b) Resolving parallel to the slope gives,

$$F = 49 \sin 5° = 4.27 \text{ N} \text{ (to 3 s.f.)}$$

(c) Using the friction law $F \leqslant \mu R$, gives,

Using **2**

$$49 \sin 5° \leqslant \mu \times 49 \cos 5°$$

$$\mu \geqslant \frac{49 \sin 5°}{49 \cos 5°}$$

$$\mu \geqslant \tan 5°$$

$$\mu \geqslant 0.0875 \text{ (to 3 s.f.)}$$

3

Worked example 2

A box of mass m kg is supported by two ropes as shown in the diagram.

The tension in the left-hand rope is 40 N.

(a) Find the tension in the right-hand rope.

(b) Find m.

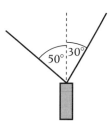

The diagram shows the forces acting on the box.

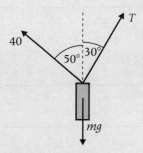

(a) Resolving horizontally gives,

$$T \sin 30° = 40 \sin 50°$$

Using **6**

$$T = \frac{40 \sin 50°}{\sin 30°} = 61.3 \text{ N} \text{ (to 3 s.f.)}$$

(b) Resolving vertically gives,

$$T \cos 30° + 40 \cos 50° = 9.8m$$

$$m = \frac{61.284 \sin 30° + 40 \cos 50°}{9.8} = 8.04 \text{ kg} \text{ (to 3 s.f.)}$$

Worked example 3

The diagram shows the forces acting on a crate, which is at rest on a rough horizontal surface. The weight of the crate is 800 N and a force of magnitude 200 N acts on the crate at an angle of 30° above the horizontal. The normal reaction acting on the crate has magnitude R and the friction force has magnitude F.

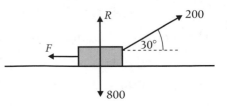

(a) Find R.

(b) Find F.

(c) The coefficient of friction between the crate and the ground is μ. Find an inequality that μ must satisfy.

(a) Resolving vertically gives,

$$R + 200 \sin 30° = 800$$

$$R = 800 - 200 \sin 30° = 700 \text{ N}$$

Using **6**

(b) Resolving horizontally gives,

$$F = 200 \cos 30° = 173 \text{ N (to 3 s.f.)}$$

(c) Using $F \leqslant \mu R$, gives,

$$200 \cos 30° \leqslant \mu \times 700$$

$$\mu \geqslant \frac{200 \cos 30°}{700}$$

$$\mu \geqslant 0.247 \text{ (to 3 s.f.)}$$

Using **2**

Worked example 4

The forces $\mathbf{F}_1 = (4\mathbf{i} + 9\mathbf{j})$ N, $\mathbf{F}_2 = (3\mathbf{i} - 7\mathbf{j})$ N and $\mathbf{F}_3$ are in equilibrium.

(a) Find $\mathbf{F}_3$.

(b) Find the magnitude of $\mathbf{F}_3$.

(c) Find the angle between $\mathbf{F}_3$ and the unit vector $\mathbf{i}$.

(a) The resultant of the three forces must be zero, which gives,

$$\mathbf{F}_1 + \mathbf{F}_2 + \mathbf{F}_3 = 0\mathbf{i} + 0\mathbf{j}$$

$$4\mathbf{i} + 9\mathbf{j} + 3\mathbf{i} - 7\mathbf{j} + \mathbf{F}_3 = 0\mathbf{i} + 0\mathbf{j}$$

$$7\mathbf{i} + 2\mathbf{j} + \mathbf{F}_3 = 0\mathbf{i} + 0\mathbf{j}$$

$$\mathbf{F}_3 = -7\mathbf{i} - 2\mathbf{j}$$

Using **6**

(b) The diagram shows the force $\mathbf{F}_3$.

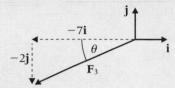

The magnitude of this force is given by,

$$\mathbf{F}_3 = \sqrt{7^2 + 2^2}$$

$$= \sqrt{53}$$

$$= 7.28 \text{ (to 3 s.f.)}$$

(c) The angle θ can be found using trigonometry.

$$\tan \theta = \frac{2}{7}$$

$$\theta = 15.9° \text{ (to 3 s.f.)}$$

The angle between the force and the unit vector $\mathbf{i}$ will be $180 - 15.9 = 164.1°$.

3

REVISION EXERCISE 3

1 A box has mass 26 kg. It is placed on a horizontal surface. Calculate the magnitude of the normal reaction force acting on the box.

2 The diagram shows the forces acting on a particle that is moving on a straight line at a constant speed.

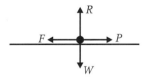

Given that $W = 300$ and $P = 250$, find R and F.

3 A car is travelling at a constant speed on a straight horizontal line. The car experiences a resistance force of magnitude 600 N. State the magnitude of the forward driving force acting on the car.

4 A sphere, which has mass 20 kg, is attached to an elastic string. The sphere is in contact with a horizontal surface, with the elastic string vertical. The tension in the string is 20 N.

   (a) Calculate the weight of the sphere.

   (b) Find the normal reaction force acting on the sphere.

5 The diagram shows the horizontal forces, in newtons, acting on a particle that is at rest on a smooth horizontal plane. The mass of the particle is 20 kg.

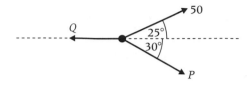

(a) Find P.

(b) Find Q.

(c) Calculate the weight of the particle.

(d) State the magnitude of the normal reaction force that the plane exerts on the particle.

6 The diagram shows the horizontal forces, in newtons, acting on a particle that is at rest on a smooth horizontal plane.

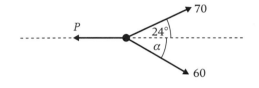

(a) By resolving perpendicular to the dashed line, find an equation that α must satisfy.

(b) Hence find α.

(c) Find P.

7 The diagram shows four forces that are in equilibrium. All the forces lie in the same plane and the magnitudes are given in Newtons.

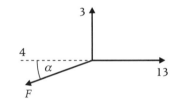

(a) Show that $\tan \alpha = \dfrac{1}{3}$.

(b) Hence find α.

(c) Find F.

8 A sack, which has weight 200 N, is supported in equilibrium by two ropes. The ropes are both at an angle of 30° to the vertical.

(a) Show that the tension in each rope is the same.

(b) Find the tension in the ropes.

9 The diagram shows the forces acting on a sign of mass 6 kg, where the sign has been modelled as a particle.

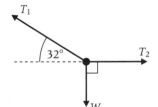

(a) Calculate the value of W.

(b) Find T_1.

(c) Find T_2.

10 A particle is placed on a rough plane, which is inclined at an angle α to the horizontal. The particle has mass 20 kg and the friction force on the particle is 45 N.

(a) Find α.

(b) Find the normal reaction force on the particle.

11 The forces **P** = $(8\mathbf{i} - 7\mathbf{j})$ N and **Q** = $(6\mathbf{i} + \mathbf{j})$ N act on a particle.

    (a) Find the resultant of **P** and **Q**.

    (b) When a third force **R** also acts the three forces are in equilibrium. Find **R**.

    (c) Find the magnitude of **R**.

    (d) Find the angle between **R** and the unit vector **i**.

12 The forces $\mathbf{F}_1 = (15\mathbf{i} - 17\mathbf{j})$ N, $\mathbf{F}_2 = (9\mathbf{i} + 14\mathbf{j})$ N, $\mathbf{F}_3 = (3\mathbf{i} + 11\mathbf{j})$ N and $\mathbf{F}_4$ are in equilibrium.

    (a) Find $\mathbf{F}_4$.

    (b) Find the magnitude of $\mathbf{F}_4$.

    (c) Find the angle between $\mathbf{F}_4$ and the unit vector **i**.

13 The diagram shows the forces acting on a particle of mass m kg.

    Given that $T_2 = 80$ N, find T_1, W and m.

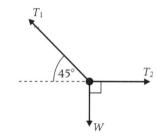

14 A box of mass m kg is supported, at rest, by two ropes as shown in the diagram.

    The tension in the left-hand rope is 600 N.

    (a) Find the tension in the right-hand rope.

    (b) Find m.

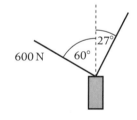

15 A particle has weight 500 N. The particle is supported by two ropes, which make angles of 40° and 30° to the vertical. Find the tension in each rope.

16 A box, which has mass 30 kg, is placed on a rough horizontal surface. A horizontal force of magnitude P Newtons is applied to the box. The coefficient of friction between the box and the surface is 0.6.

    (a) Calculate the magnitude of the normal reaction force on the box.

    (b) Calculate the maximum value of the friction force acting on the box.

    (c) State the magnitude of the friction force if:
        (i) $P = 100$ **(ii)** $P = 176.4$ **(iii)** $P = 200$

    (d) For each case above, state what happens to the box.

3

17 A block of mass 4 kg is placed on a rough horizontal surface. A horizontal force of magnitude 25 N is applied to the block. Describe what happened to the block if the coefficient of friction between the block and the surface is:

(a) 0.4   **(b)** 0.8   **(c)** 0.6

18 The diagram shows a block, of mass 12 kg, which is at rest on a rough horizontal table. The block is attached by a light string, which passes over a smooth pulley and is attached to a particle of mass 5 kg.

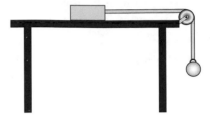

Given that the block is on the point of sliding, find the coefficient of friction between the block and the table.

19 The diagram shows four forces, in Newtons.

These forces all act in the same plane. Find the magnitude of the resultant of these four forces.

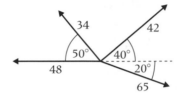

20 The diagram shows four forces, which are in equilibrium and that lie in the same plane.

(a) Find F.

(b) Find α.

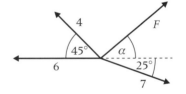

21 A box has mass 30 kg. It is at rest on a rough horizontal surface when a child tries to push the box by exerting a force, of magnitude 80 N, as shown in the diagram.

Model the box as a particle. The coefficient of friction between the box and the surface is 0.7.

(a) Find the normal reaction force acting on the box.

(b) Determine whether or not the child is able to move the box.

22 A particle, of mass 2 kg, is at rest on a slope inclined at an angle of 70° to the horizontal. The particle is held at rest by a string that is parallel to the slope, as shown in the diagram.

(a) Find the tension in the string if the slope is smooth.

(b) Find the friction force acting on the particle if the tension in the string is 6 N.

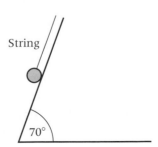

23 The diagram shows a small sphere that is at rest on a slope, inclined at an angle of 40° to the horizontal. A vertical string is attached to the particle, as shown in the diagram. The mass of the sphere is 5 kg and the tension in the string is 20 N.

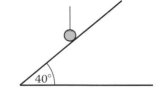

  (a) Find the magnitude of the normal reaction force acting on the sphere.

  (b) Find the magnitude of the friction force acting on the sphere.

24 The diagram shows the forces acting on a crate, of mass 100 kg, at rest on a rough horizontal surface. A force of magnitude T N acts on the crate, at an angle of 30° above the horizontal. The normal reaction acting on the crate has magnitude R and the friction force has magnitude F. The coefficient of friction between the crate and the surface is 0.5 and the crate is on the point of sliding.

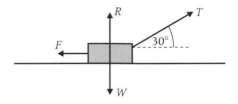

  (a) Express R in terms of T.

  (b) Find T.

25 A particle, of mass 4 kg, is on a rough plane, which is inclined at an angle of 60° to the horizontal. The particle is held at rest by a horizontal force, of magnitude 40 N, as shown in the diagram.

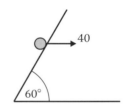

  (a) By resolving perpendicular to the plane, find the magnitude of the normal reaction force acting on the particle.

  (b) Find the magnitude of the friction force acting on the particle.

26 A force of magnitude 30 newtons acts on a particle, as shown in the diagram. The mass of the particle is 10 kg and it remains at rest on a rough plane inclined at 15° to the horizontal.

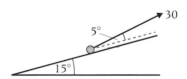

  (a) Find the magnitude of the normal reaction force acting on the particle.

  (b) Find the magnitude of the friction force acting on the particle and state the direction in which it acts.

  (c) Given that the particle is on the point of sliding, find the coefficient of friction between the particle and the plane.

27 Four forces, $\begin{bmatrix} 6 \\ 0 \end{bmatrix}$ Newtons, $\begin{bmatrix} 0 \\ 4 \end{bmatrix}$ Newtons, $\begin{bmatrix} -3 \\ -4.5 \end{bmatrix}$ Newtons and $\begin{bmatrix} 3 \\ -2 \end{bmatrix}$ Newtons, act on a particle.

  (a) Express the resultant, **F** Newtons, of these four forces as a column vector.

  (b) Find the magnitude of **F**.   [A]

28 A particle is held at rest on a smooth slope by a horizontal force of magnitude 8 Newtons, as shown in the diagram. The slope is at an angle of 60° to the horizontal.

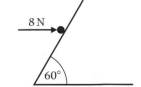

  (a) Draw a diagram to show the forces acting on the particle.

  (b) By resolving horizontally show that the magnitude of the normal reaction force acting on the particle is approximately 9.24 Newtons.

  (c) Find the mass of the particle, giving your answer to two significant figures. [A]

29 Two light, inextensible strings are attached to a particle of mass m kg. Each string passes over a fixed, smooth, light pulley. The other end of one string is attached to a particle of mass 4 kg. The other end of the second string is attached to a particle of mass 3 kg. The diagram shows the system in its equilibrium position. The angles marked on the diagram are between the strings and the vertical.

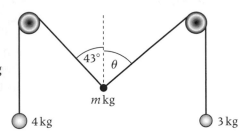

  (a) Calculate the tension in each of the strings.

  (b) Show that $\theta = 65.4°$, correct to three significant figures.

  (c) Find m. [A]

Test yourself	What to review
	If your answer is incorrect:

1 A box has weight 50 N. It is placed on a rough plane inclined 22° to the horizontal, where it remains at rest.
  (a) Find the magnitude of the normal reaction force acting on the box.
  (b) Find the magnitude of the friction force acting on the box.
  (c) The coefficient of friction between the box and the plane is μ. Find an inequality that μ must satisfy.
	See pp 24–26 Examples 1 and 3 or review Advancing Maths for AQA M1 pp 77–79

2 The diagram shows three forces that are in equilibrium. All the forces lie in the same plane and the magnitudes are given in Newtons. 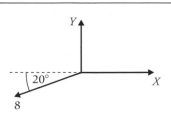 Find X and Y.	See p 25 Example 2 or review Advancing Maths for AQA M1 pp 66–68
3 The forces $\mathbf{F}_1 = (13\mathbf{i} - 6\mathbf{j})$ N, $\mathbf{F}_2 = (19\mathbf{i} + a\mathbf{j})$ N and $\mathbf{F}_3 = (b\mathbf{i} + 31\mathbf{j})$ N are in equilibrium. Find a and b.	See p 26 Example 4 or review Advancing Maths for AQA M1 pp 66–68

4 As a child, of mass 50 kg, moves down a slide he travels at a constant speed. The slide is inclined at an angle of 45° to the horizontal. The coefficient of friction between the slide and the child is 0.4.

   (a) Find the magnitude of the normal reaction force acting on the child.

   (b) Hence find the magnitude of the friction force acting on the child.

   (c) Determine the magnitude of the air resistance force acting on the child.

Review Adancing Maths for AQA M1 pp 77–79

3

5 Two ropes are attached to a load of mass 500 kg. The ropes make angles of 30° and 45° to the vertical as shown in the diagram. The tensions in these ropes are T_1 and T_2 Newtons. The load is also supported by a vertical spring.

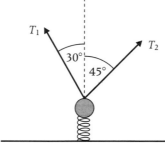

See p 25 Example 2 or review Advancing Maths for AQA M1 pp 66–68

The system is in equilibrium and $T_1 = 200$.

   (a) Show that $T_2 = 141$, correct to three significant figures.

   (b) Find the force that the spring exerts on the load.   [A]

Test yourself **ANSWERS**

5 **(b)** 4630 N

4 **(a)** 346 N   **(b)** 139 N   **(c)** 208 N

3 $a = -25$, $b = -32$

2 $X = 7.52$, $Y = 2.74$

1 **(a)** 46.4 N   **(b)** 18.7 N   **(c)** $\mu \geqslant 0.404$

CHAPTER 4

Newton's second law

Key points to remember

1 Newton's second law states the resultant force on a body is equal to the product of its mass and acceleration.

    (i)   For motion in a straight line this can be expressed as $F = ma$.

    (ii) For motion in two or three dimensions the law should be expressed using vectors as
       $\mathbf{F} = m\mathbf{a}$.

Note that the law should be applied to the resultant of all the forces acting.

Worked example 1

A car, of mass 1200 kg, accelerates at 2 m s^{-2} on a straight horizontal road. It experiences a resistance force of magnitude 750 N. Find the magnitude of the driving force that acts on the car.

The diagram shows the forces acting on the car. Note that the car has been modelled as a particle.

The driving force has been labelled D.

Note that the vertical forces must balance as the car is travelling horizontally.

Consider the horizontal forces to find the resultant force.

Resultant force $= D - 750$

Now apply Newton's second law, $F = ma$.

$D - 750 = 1200 \times 2$
$D - 750 = 2400$
$D = 2400 + 750$
$= 3150 \text{ N}$

34522

Using **1** (i)

Worked example 2

The forces $(9\mathbf{i} + 8\mathbf{j})\,N$, $(7\mathbf{i} - 2\mathbf{j})\,N$ and $(-4\mathbf{i} + 7\mathbf{j})\,N$ act on a particle of mass 5 kg.

(a) Find the acceleration of the particle.

(b) Find the magnitude of the acceleration of the particle.

(c) Find the angle between the unit vector $\mathbf{i}$ and the acceleration.

(a) First find the resultant force on the particle.

$$\text{Resultant force} = 9\mathbf{i} + 8\mathbf{j} + 7\mathbf{i} - 2\mathbf{j} - 4\mathbf{i} + 7\mathbf{j}$$
$$= 12\mathbf{i} + 13\mathbf{j}$$

Then apply Newton's second law, $\mathbf{F} = m\mathbf{a}$.

$$12\mathbf{i} + 13\mathbf{j} = 5\mathbf{a}$$
$$\mathbf{a} = \frac{12\mathbf{i} + 13\mathbf{j}}{5} = 2.4\mathbf{i} + 2.6\mathbf{j}$$

Using ■ (ii)

(b) Now find the magnitude of the acceleration vector.

$$a = \sqrt{2.4^2 + 2.6^2}$$
$$= 3.54 \text{ m s}^{-2} \text{ (to 3 s.f.)}$$

4

(c) The diagram shows the angle between the acceleration and the unit vector $\mathbf{i}$.

$$\tan \alpha = \frac{2.6}{2.4}$$

$$\alpha = \tan^{-1}\left(\frac{2.6}{2.4}\right)$$

$$= 47.3° \text{ (to 3 s.f.)}$$

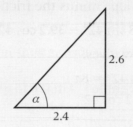

Worked example 3

A wooden block of mass 8 kg is placed on a slope inclined at 42° to the horizontal.

(a) If the slope is smooth, find the acceleration of the block.

(b) Find the magnitude of the normal reaction force that acts on the block.

(c) If the coefficient of friction between the block and the slope is 0.5, find the acceleration of the particle.

(a) The diagram shows the forces acting on the block.

The resultant force will be the component of the weight parallel to the slope.

$$\text{Resultant force} = 8 \times 9.8 \sin 42°$$

Using Newton's second law gives,

$$8 \times 9.8 \sin 42° = 8a$$
$$a = 9.8 \sin 42°$$
$$= 6.56 \text{ m s}^{-2} \text{ (to 3 s.f.)}$$

Using ■ (i)

7 A ball of mass 0.2 kg is released from rest. As it falls its acceleration is 9.2 m s^{-2}. Find the magnitude of the resistance force that is acting on the ball as it falls.

8 A wooden block, of mass 2 kg, is set into motion, so that it slides in a straight line across a rough horizontal table. The coefficient of friction between the block and the table is 0.3.

(a) Find the magnitude of the friction force acting on the block.

(b) Find the acceleration of the block.

(c) Given that the block was initially moving at 8 m s^{-1}, find the distance that it travels before it comes to rest.

9 A crane is used to lift a load vertically. A cable is attached to the load, which has mass 450 kg. The load rises from rest with a constant acceleration, reaching a speed of 2 m s^{-1} after 5 seconds.

(a) Find the acceleration of the load.

(b) Find the tension in the cable.

10 A motorcycle, of mass 350 kg, travels on a horizontal road. As it moves it experiences a driving force of magnitude 1400 N. It also experiences a resistance force of magnitude 28v newtons, where v m s^{-1} is the speed of the motorcyclist. Find:

(a) the acceleration of the motorbike when travelling at 10 m s^{-1};

(b) the speed at which its acceleration is 2 m s^{-2};

(c) the maximum speed of the motorcycle.

11 A rope is attached to a car of mass 1200 kg. Assume that there is no resistance to the motion of the car. Find the acceleration of the car if the tension in the rope is 750 N and the rope is:

(a) horizontal;

(b) at an angle of 30° to the horizontal.

12 The diagram shows the forces acting on a particle of mass 50 kg, as it moves on a rough horizontal surface. The coefficient of friction between the surface and the particle is 0.4.

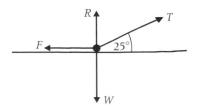

(a) Calculate the magnitude of the weight, W, of the particle.

(b) Given that $T = 200$, find R.

(c) Find the magnitude, F, of the friction force.

(d) Find the acceleration of the particle.

13 The diagram shows the forces that act on a particle of mass 4 kg. The particle moves along a straight line.

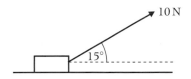

(a) Find the angle α.

(b) Find the acceleration of the particle.

14 A force, $(5\mathbf{i} - 8\mathbf{j})$ N acts on particle of mass m. The acceleration of the particle is $(2\mathbf{i} - f\mathbf{j})$ m s^{-2}.

(a) Find m.

(b) Find f.

15 The forces in newtons, listed below, act on a particle of mass 5 kg.

$$4\mathbf{i} - 7\mathbf{j} \qquad 3\mathbf{i} + 5\mathbf{j} \qquad -6\mathbf{i} + 9\mathbf{j}$$

(a) Find the resultant force on the particle.

(b) Find the acceleration of the particle.

(c) Find the magnitude of the acceleration of the particle.

16 A particle has mass 2 kg and accelerates in the $\mathbf{j}$ direction. The forces in newtons, listed below, act on the particle.

$$3\mathbf{i} + 9\mathbf{j} \qquad 4\mathbf{i} - 7\mathbf{j} \qquad -15\mathbf{i} + \mathbf{j} \qquad F(\mathbf{i} + \mathbf{j})$$

(a) Find F.

(b) Find the acceleration of the particle.

17 The diagram shows a rope, which is attached to a box of mass 8 kg. The box is on a horizontal surface.

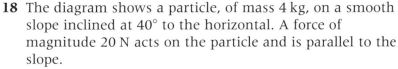

Model the box as a particle.

(a) Find the acceleration of the box if the surface is smooth.

(b) Find the magnitude of the normal reaction force acting on the box.

(c) Find the acceleration of the box if the coefficient of friction between the surface and the box is 0.1.

18 The diagram shows a particle, of mass 4 kg, on a smooth slope inclined at 40° to the horizontal. A force of magnitude 20 N acts on the particle and is parallel to the slope.

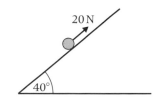

(a) Find the resultant force on the particle.

(b) Find the acceleration of the particle.

19 A car, of mass 1000 kg, rolls down a slope inclined at an angle of 5° to the horizontal. Find the acceleration of the car if there is no resistance to its motion.

20 A van, of mass 2500 kg, is subject to a resistance force of 500 N as it moves. If it rolls down a slope inclined at 6° to the horizontal, find the acceleration of the van.

21 A particle slides down a smooth slope inclined at an angle α to the horizontal. Given that the acceleration of the particle is 3 m s^{-2}. Find α.

22 A sledge, of mass 5 kg, is placed on a slope, which is inclined at 40° to the horizontal. The coefficient of friction between the sledge and the slope is 0.2.
   (a) Find the magnitude of the normal reaction force acting on the sledge.
   (b) Find the magnitude of the friction force that acts on the sledge when it moves.
   (a) Find the acceleration of the sledge.

23 A particle, of mass 20 kg, slides down a rough inclined plane. The angle between the plane and the horizontal is 40°. The coefficient of friction between the particle and the slope is μ.
   (a) Find μ if the particle slides at a constant speed.
   (b) Find μ if the acceleration of the particle is 0.3 m s^{-2}.
   (c) State one assumption that you have made to obtain your answers.

24 A lorry, of mass 12 tonnes, travels in a straight line up a slope inclined at 3° to the horizontal. The lorry accelerates at 0.05 m s^{-2}. A forward driving force acts on the lorry in the direction of motion.
   (a) Assume that there is no resistance to the motion of the lorry.
      (i) Find the magnitude of the driving force.
      (ii) The driving force is removed. Calculate the new acceleration of the lorry in this situation.
   (b) A revised model for the motion of the lorry assumes that there is a resistance force of 5000 N opposing the motion of the lorry. Find the driving force acting on the lorry as it accelerates at 0.05 m s^{-2} up the slope.

25 A skier slides down a slope inclined at 40° to the horizontal. The coefficient of friction between the skier and the slope is 0.2. The mass of the skier is 65 kg.
   (a) Find the magnitude of the friction force acting on the skier.

(b) Find the acceleration of the skier, assuming that there is no air resistance.

(c) If the acceleration of the skier is 10% less than the value you obtained in part **(b)**, calculate the magnitude of the air resistance force acting on the skier.

26 A rope is used to pull a sledge, of mass 8 kg, up a snow covered slope. The rope is pulled so that it remains parallel to the slope. The slope is inclined at 45° to the horizontal. As it moves the sledge accelerates at 0.1 m s^{-2}. The coefficient of friction between the slope and the sledge is 0.4. Find the tension in the rope.

27 A child kicks a small toy brick in a straight line across a horizontal floor. The brick initially moves at 3.5 m s^{-1} and comes to rest in a distance of 2.5 metres.

(a) Show that the magnitude of the retardation of the brick is 2.45 m s^{-2}.

(b) The mass of the brick is 0.2 kg.
   (i) Find the magnitude of the frictional force acting on the brick.
   (ii) Find the coefficient of friction between the brick and the floor. [A]

28 A box, of mass 20 kg, is initially at rest on a rough horizontal surface. A horizontal force of magnitude P newtons is applied to the box. The coefficient of friction between the box and the surface is 0.3.

(a) State the magnitude of the normal reaction force acting on the box.

(b) Find the magnitude of the friction force that acts on the box if:
   (i) $P = 80$ **(ii)** $P = 40$.

(c) Find the value of P when the box is accelerating at 0.8 m s^{-2}.

(d) When the box reaches a speed of 6 m s^{-1}, the horizontal force P is removed. Find the distance that the box travels after the force P is removed. [A]

29 A particle of mass 4 kg moves on a smooth horizontal plane. It is initially at rest at the origin. A force $\mathbf{F} = (8\mathbf{i} - 12\mathbf{j})$ N acts on the particle for 20 seconds. The unit vectors $\mathbf{i}$ and $\mathbf{j}$ are perpendicular and lie in the horizontal plane.

(a) **(i)** Find the acceleration of the particle.
   (ii) Find the velocity of the particle at the end of the 20 second period.
   (iii) Find the position of the particle at the end of the 20 second period.

(b) At the end of the 20 second period, the force **F** is removed. Find the distance of the particle from the origin after the particle has been in motion for a total of 45 seconds. [A]

30 A block of wood has mass 4 kg. It is placed on a rough horizontal surface and is pulled by a horizontal string. The coefficient of friction between the block and the surface is 0.4.

(a) Draw a diagram to show the forces acting on the block.

(b) Calculate the magnitude of the normal reaction force acting on the block.

(c) If the acceleration of the block is 2 m s^{-2}, find the tension in the string.

(d) If the tension in the string is 20 N, find the acceleration of the block. [A]

31 A mountain railway train moves on a straight track. The mass of the train and its passengers is 1000 kg. During its motion the train moves under the action of a variable propulsive force, P newtons, and a constant resistance force of R newtons.

(a) During the first stage of its motion, the train moves horizontally with acceleration 0.25 m s^{-2}. In this stage, the value of P is 1200.

Show that $R = 950$.

(b) During the second stage of its motion, the train moves up a slope inclined at an angle α to the horizontal, where $\sin \alpha = 0.1$.

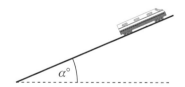

In this stage, the value of $P = 2100$, and the value of R remains at 950. Find the acceleration of the train. [A]

Test yourself	**What to review**
	If your answer is incorrect:
1 The tension in a lift cable is 3000 N. The lift and its passengers have a total mass of 290 kg. Find the acceleration of the lift.	See p 34 Example 2 or review *Advancing Maths for AQA M1* pp 87–89
2 An object, of mass 10 kg is allowed to fall vertically from rest. It experiences a resistance force of 34 N as it falls. Find the acceleration of the object.	See p 34 Example 1 or review *Advancing Maths for AQA M1* pp 87–89

3 The forces in newtons, listed below, act on a particle of mass 7 kg.

$$12\mathbf{i} - 5\mathbf{j} \qquad -3\mathbf{i} + 8\mathbf{j} \qquad p\mathbf{i} + q\mathbf{j}$$

The acceleration of the particle is $(2\mathbf{i} + 3\mathbf{j})$ m s^{-2}. Find the values of p and q.

See p 34 Example 2 or review Advancing Maths for AQA M1 pp 87–89

4 A particle, of mass 2 kg, is placed on a smooth slope inclined at 38° to the horizontal. Find the acceleration of the particle.

See p 35 Example 3 or review Advancing Maths for AQA M1 pp 87–89

5 The diagram shows the force applied to a block of mass 15 kg, which is on a rough horizontal surface. The coefficient of friction between the block and the surface is 0.4.

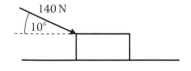

See p 35 Example 3 or review Advancing Maths for AQA M1 pp 87–89

4

(a) Find the magnitude of the normal reaction force acting on the block.

(b) Find the acceleration of the block.

Test yourself ANSWERS

5 (a) 171 N   **(b)** 4.62 m s^{-2}

4 6.03 m s^{-2}

3 $p = 5$, $q = 18$

2 6.4 m s^{-2}

1 0.545 m s^{-2}

CHAPTER 5

Connected particles

Key points to remember

1 You should always form an equation of motion for each particle, by applying Newton's second law. When you have two equations, solve them using the standard techniques for simultaneous equations. It is important to note that the examiner will be looking for two equations of motion.

Worked example 1

Two particles are connected by a light string that passes over a smooth peg, as shown in the diagram.

The particles have masses of 5 kg and 3 kg.

(a) Find the acceleration of the particles.

(b) Find the tension in the string.

(a) First consider each particle separately. The diagrams show the forces on each particle.

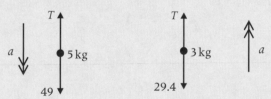

For the 5 kg particle, Newton's second law gives,

$$49 - T = 5a$$

For the 3 kg particle, Newton's second law gives,

$$T - 29.4 = 3a \text{ or } T = 3a + 29.4$$

Now substitute for T in the first equation to give,

$$49 - (3a + 29.4) = 5a$$

$$49 - 3a - 29.4 = 5a$$

$$19.6 = 8a$$

$$a = \frac{19.6}{8} = 2.45 \text{ m s}^{-2}$$

Using **1**

(b) To find the tension substitute $a = 2.45$ into the expression $T = 3a + 29.4$, which gives,

$$T = 3 \times 2.45 + 29.4$$

$$= 36.8 \text{ N (to 3 s.f.)}$$

Worked example 2

A wooden block, of mass 2 kg, is at rest on a rough, horizontal table. A light string is attached to the block. The string passes over a smooth, light pulley and the other end is attached to a particle of mass 3 kg. The coefficient of friction between the block and the table is 0.3.

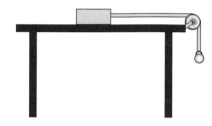

Find the acceleration of the block and the tension in the string.

The diagrams show the forces acting on the block and on the particle.

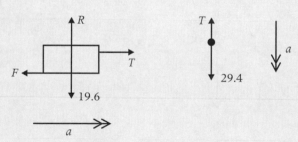

First consider the 3 kg particle. Applying Newton's second law gives,

$$29.4 - T = 3a$$

Considering the vertical forces on the 2 kg block gives,

$$R = 19.6$$

This can then be used to find the magnitude of the friction force acting on the block, by using $F = \mu R$.

$$F = 0.3 \times 19.6 = 5.88$$

Now apply Newton's second law to the block.

$$T - 5.88 = 2a \quad \text{or} \quad T = 2a + 5.88$$

Using **1**

5

This can then be used to eliminate T from the equation $29.4 - T = 3a$ which applies to the particle.

$$29.4 - (2a + 5.88) = 3a$$

$$29.4 - 2a - 5.88 = 3a$$

$$23.52 = 5a$$

$$a = \frac{23.52}{5} = 4.70 \text{ m s}^{-2} \text{ (to 3 s.f.)}$$

Now substitute this acceleration into the equation $T = 2a + 5.88$ to find the tension.

$$T = 2 \times 4.70 + 5.88 = 15.3 \text{ N (to 3 s.f.)}$$

Worked example 3

Two particles are connected by a light string that passes over a smooth, light pulley. One particle has mass 20 kg and is on a smooth slope inclined at 30° to the horizontal. The other end of the string is attached to a particle, of mass 30 kg, which is directly below the pulley. The system is shown in the diagram.

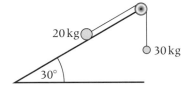

(a) Find the acceleration of the particles.

(b) Find the tension in the string.

(a) The diagrams below show the forces acting on each particle.

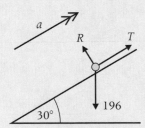 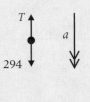

Applying Newton's second law to the particle on the slope gives,

$$T - 196 \sin 30° = 20a$$

$$T - 98 = 20a$$

or

$$T = 20a + 98$$

Applying Newton's second law to the particle below the pulley gives,

$$294 - T = 30a$$

Using **I**

Now the result $T = 20a + 98$, from the first equation, can be used to substitute for T in the second equation to give,

$$294 - (20a + 98) = 30a$$

$$294 - 20a - 98 = 30a$$

$$196 = 50a$$

$$a = \frac{196}{50} = 3.92 \text{ m s}^{-2}$$

(b) This value for a can now be substituted into the equation $T = 20a + 98$ to give the value of T.

$$T = 20 \times 3.92 + 98$$

$$= 176 \text{ N (to 3 s.f.)}$$

5

Worked example 4

A car of mass 1400 kg tows a caravan of mass 1100 kg, on a straight horizontal road. The car experiences a resistance force of 500 N and the caravan experiences a resistance force of 800 N. Given that the car and caravan are accelerating at 0.8 m s^{-2}, find:

(a) the forward force that acts on the car;

(b) the force that the car exerts on the caravan.

(a) Model the car and caravan as a single particle. The diagram shows the horizontal forces that are acting, where D is the driving force exerted by the car.

First find the resultant force on the car and caravan.

$$\text{Resultant force} = D - 800 - 500$$

$$= D - 1300$$

Applying Newton's second law gives,

Using **1**

$$D - 1300 = (1400 + 1100) \times 0.8$$

$$D - 1300 = 2000$$

$$D = 3300 \text{ N}$$

(b) Now consider the forces acting on the caravan. There will be the 800 N resistance force and a forward force T N as shown in the diagram.

First find the resultant force on the caravan.

Resultant force $= T - 800$

Applying Newton's second law gives,

$T - 800 = 1100 \times 0.8$

$T - 800 = 880$

$T = 1680 \, \text{N}$

REVISION EXERCISE 5

1 Two particles are connected by a light string that passes over a smooth peg, as shown in the diagram.

The particles have masses of 4 kg and 1 kg.

(a) Find the acceleration of the particles.

(b) Find the tension in the string.

4 kg   1 kg

2 Two particles are connected by a light string that passes over a smooth peg. One particle has mass 3 kg and the other mass 7 kg. The system is held at rest and the rope taut. The particles are released from the same height and the two particles move vertically.

(a) Find the acceleration of the particles.

(b) Find the tension in the string.

3 Two particles are connected by a light string that passes over a smooth, light pulley. The particles are released from rest with the rope taut and move vertically. Find the acceleration of the particles if their masses are:

(a) 2 kg and 12 kg;

(b) 12 kg and 8 kg;

(c) 20 kg and 30 kg.

4 Two particles are connected by a light string that passes over a smooth peg. The particles are released from rest with the rope taut and move vertically. Find the tension in the string if the masses of the particles are:

(a) 3 kg and 11 kg;

(b) 20 kg and 8 kg;

(c) 100 kg and 5 kg.

5 Two particles are connected by a light string that passes over a smooth, light pulley, as shown in the diagram.

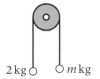

One particle has a mass of 2 kg and the other has a mass of m kg. The 2 kg particle accelerates upwards at 4.2 m s^{-2}.

(a) Find the tension in the string.

(b) Find m.

6 Two particles are connected by a light string that passes over a smooth peg. The particles are released from rest and move vertically with the string taut. After 4 seconds the particles are travelling at 6 m s^{-1}.

(a) Find the acceleration of the particles.

(b) The mass of the particle which is moving downwards is 5 kg. Find the tension in the string.

(c) Find the mass of the other particle.

7 Two particles have masses of $25m$ kg and $24m$ kg. They are connected by a light string that passes over a pulley. The particles are released from rest and move vertically with the string taut.

(a) State two assumptions that it would be appropriate to make about the pulley.

(b) State one assumption that it would be appropriate to make about the particles.

(c) Find the acceleration of the particles in terms of g.

(d) Find the tension in the string in terms of m and g.

8 Two particles of mass m kg and M kg are connected by a light string that passes over a smooth peg. The particles are released from rest. They then move vertically with the string taut. Assume that $M > m$.

(a) Find the acceleration of the particles in terms of m, M and g.

(b) Find the tension in the string in terms of m, M and g.

9 Two particles have masses of $\dfrac{4m}{5}$ kg and $\dfrac{6m}{5}$ kg. They are connected by a light string that passes over a smooth peg. The particles are released from rest and move vertically with the string taut.

(a) Find the acceleration of the particles in terms of g.

(b) Find the tension in the string in terms of m and g.

10 A particle, of mass 4 kg, is placed on a rough table at a distance of 1 metre from a smooth hole. A light string is attached to the particle and passes through the hole. A second particle, of mass 10 kg, is attached to the other end of the string and is hanging vertically beneath the hole when the first particle is released. The coefficient of friction between the particle and the table is 0.3.

(a) Find the magnitude of the friction force between the particle and the table.

(b) Find the acceleration of the particles.

(c) Find the tension in the string.

11 A sphere, of mass 5 kg, is placed on a smooth horizontal surface and connected to a particle, of mass 9 kg, by a light string. The string passes over a smooth light pulley as shown in the diagram.

The sphere is released from rest.

(a) Find the acceleration of the sphere.

(b) Find the time it takes the sphere to travel 50 cm.

(c) Find the tension in the string.

12 The diagram shows a block, of mass M, that is attached to a particle, of mass 7 kg, by a light string that passes over a smooth, light pulley. The block is on a rough surface and the coefficient of friction between the block and the surface is 0.4. The block starts at rest and reaches a speed of 0.6 m s^{-1} after it has travelled 0.5 metres.

(a) Find the acceleration of the block.

(b) Find the tension in the string.

(c) Find M.

13 The diagram shows a block, of mass M, which rests on a rough horizontal surface and is connected to a particle of mass m by a light string, as shown.

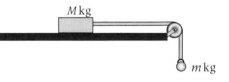
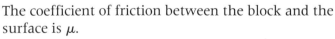
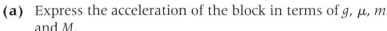

The coefficient of friction between the block and the surface is μ.

(a) Express the acceleration of the block in terms of g, μ, m and M.

(b) Express the tension in the string in terms of g, μ, m and M.

14 The diagram shows a block, of mass 6 kg, that is on a smooth horizontal surface. It is attached to particles of mass 5 kg and 3 kg, as shown.

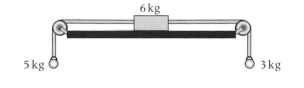

(a) Find the acceleration of the block.

(b) Find the tension in the left-hand string.

(c) Find the tension in the right-hand string.

15 A particle, of mass 3 kg, is placed on a rough table at a distance of 2 metres from a smooth hole. A light string is attached to the particle and passes through the hole. A second particle, of mass 7 kg, is attached to the other end of the string and is hanging vertically beneath the hole when the first particle is released. The coefficient of friction between the particle and the table is μ. The system is released from rest and it takes the 3 kg particle 1 second to reach the hole.

(a) Find the acceleration of the particles.

(b) Find the tension in the string.

(c) Find μ.

16 The diagram shows a wooden block, of mass 8 kg, which is attached to a second particle, of mass 6 kg, by a light string. The string passes over a smooth, light pulley. The coefficient of friction between the block and the table is μ. The block is released from rest and travels 0.6 metres in 0.8 seconds.

(a) Find the acceleration of the block.

(b) Find the tension in the string.

(c) Find μ.

17 A toy train set has an engine of mass 0.8 kg and two carriages of mass 0.5 kg each. A resistance force of 2 N acts on the engine and resistance forces of 1.5 N act on each carriage. A forward force of 6 N acts on the engine.

(a) Find the acceleration of the train.

(b) Find the force that the engine exerts on the first carriage.

(c) Find the force that the first carriage exerts on the second carriage.

18 A tractor, of mass 4000 kg, tows a trailer, of mass 8000 kg. A resistance force of 300 N acts on the trailer and a resistance force of 200 N acts on the tractor. The tractor and trailer accelerate at 0.04 m s^{-2}.

(a) Find the force that the trailer exerts on the tractor.

(b) Find the magnitude of the forward force that acts on the tractor.

5

19 A van is tied to a skip that is at rest on rough horizontal ground. The mass of the skip is 200 kg and the coefficient of friction between the skip and the ground is 0.7. The rope is horizontal and the mass of the van is 1800 kg. The van and the skip accelerate at 0.02 m s⁻².

(a) Find the friction force that acts on the skip.

(b) Find the tension in the rope.

(c) Find the forward driving force that acts on the van.

20 A car of mass 1600 kg tows a caravan of mass 900 kg. The car experiences a resistance force of 500 N and the caravan experiences a resistance force of 800 N. Given that the car and caravan are accelerating at 0.8 m s⁻², find the forward force that acts on the car and the force that the car exerts on the caravan, if they are moving:

(a) on a horizontal road;

(b) up a slope inclined at 4° to the horizontal.

21 A smooth slope is inclined at 60° to the horizontal. Two connected particles have masses of 10 kg and 4 kg. The particles are connected by a light string that passes over a smooth peg. The diagram shows the initial positions of the two particles.

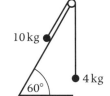

(a) Find the acceleration of the particles.

(b) Find the tension in the string.

(c) Find the time that it would take the 10 kg particle to travel 0.8 metres from rest.

22 The diagram shows two particles connected by a light string that passes over a smooth peg.

The slope is smooth and the 2 kg particle accelerates up the slope at 2 m s⁻².

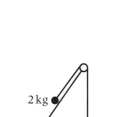

(a) Find the tension in the string.

(b) Find m.

23 The diagram shows two particles connected by a light string that passes over a smooth peg. One particle has mass m kg and is on a smooth slope inclined at 18° to the horizontal. The other end of the string is attached to a particle, of mass 3 kg, which is directly below the peg.

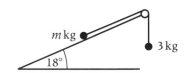

When the particles are released from rest they accelerate at 0.9 m s⁻², with the 3 kg particle moving downwards.

(a) Find the tension in the string.

(b) Find m.

24 Two particles, in the diagram, are connected by a light string that passes over a smooth peg. One particle has mass 10 kg and the other has mass 6 kg. The 10 kg particle is on a smooth slope inclined at $\alpha°$ to the horizontal. The 6 kg is directly below the peg.

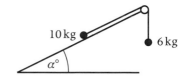

When the particles are released from rest, with the string taut, they accelerate at 1 m s^{-2}, with the 10 kg particle moving down the slope.

(a) Find α.

(b) Find the tension in the string.

25 Two particles are connected by a light string that passes over a smooth peg. One particle has mass 2 kg and is on a rough slope inclined at 40° to the horizontal. The other end of the string is attached to a particle, of mass 3 kg, which is directly below the pulley. The system is shown in the diagram.

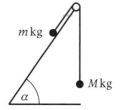

5

The coefficient of friction between the particle and the slope is 0.2. The particles are released from rest with the string taut.

(a) Calculate the magnitude of the friction force acting on the 2 kg particle.

(b) Find the acceleration of the particles.

(c) Find the tension in the string.

26 A smooth slope is inclined at $\alpha°$ to the horizontal. Two connected particles have masses of m kg and M kg. The particles are connected by a light string that passes over a smooth peg. The diagram shows the initial positions of the two particles, which are released from rest with the string taut.

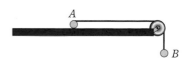

Find the acceleration of the particles and the tension in the string, in terms of g, m, M and α, given that the M kg particle rises.

27 The diagram shows two particles, A and B, that are connected by a light inextensible string. Particle A has mass 8 kg and is on a rough horizontal surface. Particle B has mass 7 kg and is attached to the other end of string as shown in the diagram. The string passes over a smooth light pulley.

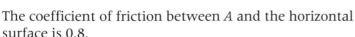

The coefficient of friction between A and the horizontal surface is 0.8.

(a) Find the maximum magnitude of the friction force between A and the surface.

(b) By forming an equation of motion for each particle, show that the magnitude of the acceleration of each particle is 0.392 m s^{-2}. [A]

28 Two particles are connected by a string, which passes over a pulley. Model the string as light and inextensible. The particles have masses of 2 kg and 5 kg. The particles are released from rest.

(a) State one modelling assumption that you should make about the pulley.

(b) By forming an equation of motion for each particle, show that the magnitude of the acceleration of each particle is 4.2 m s^{-2}.

(c) Find the tension in the string. [A]

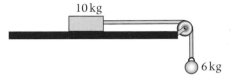

29 A block, of mass 10 kg, rests on a rough horizontal surface. It is connected by a light, inextensible string to a particle of mass 6 kg. The string passes over a light, smooth pulley, so that the string hangs vertically, as shown in the diagram.

Model the block as a particle.

(a) The system is released from rest and the block travels 0.5 metres in 2 seconds. Find the acceleration of the system.

(b) Show that the tension in the string is 57.3 N.

(c) Find the coefficient of friction between the block and the plane. [A]

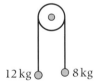

30 A light, inextensible string has a particle of mass 8 kg attached to one end and a particle of mass 12 kg attached to the other end. The string passes over a smooth, light pulley. The particles are released from rest with the string taut and vertical on each side of the pulley. The diagram shows the pulley and the particles.

(a) By forming an equation of motion for each particle, show that the acceleration of the particles is 1.96 m s^{-2}.

(b) Find the tension in the string.

(c) Find the time that it takes for the particles to reach a speed of 7 m s^{-1}. [A]

Test yourself	**What to review**

If your answer is incorrect:

1 A light string passes over a smooth peg. The string is attached to particles of mass 4 kg and 6.5 kg. The particles are released from rest with the string taut. The particles then move vertically.

   (a) Find the acceleration of the particles.

   (b) Find the time that it takes for the particles to reach a speed of 7 ms^{-1}.

   (c) Find the distance that each particle moves in this time.

See p 44 Example 1 or review *Advancing Maths for AQA M1* p 100

---

2 A wooden block, of mass 18 kg, is at rest on a rough, horizontal table. A light string is attached to the block. The string passes over a smooth, light pulley and the other end is attached to a particle of mass 10 kg. The coefficient of friction between the block and the table is 0.2.

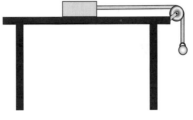

   (a) Find the magnitude of the friction force acting on the block.

   (b) Find the acceleration of the block.

   (c) Find the tension in the string.

   (d) Find the speed of the block when it has travelled 1.2 metres.

See p 45 Example 2 or review *Advancing Maths for AQA M1* p 101

---

3 A car of mass 1200 kg tows a trailer of mass 200 kg, on a straight horizontal road. The car experiences a resistance force of 400 N and the trailer experiences a resistance force of 50 N. Given that the car and trailer accelerate at 0.8 m s^{-2}, find:

   (a) the forward force that acts on the car;

   (b) the force that the car exerts on the trailer.

See p 47 Example 4 or review *Advancing Maths for AQA M1* p 103

---

4 Two particles are connected by a light string that passes over a smooth, light pulley. One particle has mass 10 kg and is on a smooth slope inclined at 25° to the horizontal. The other end of the string is attached to a particle, of mass 30 kg, which is directly below the pulley. The system is shown in the diagram.

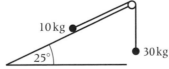

   (a) Find the acceleration of the particles.

   (b) Find the tension in the string.

See p 46 Example 3 or review *Advancing Maths for AQA M1* pp 102–103

Test yourself (continued)

What to review

If your answer is incorrect:

5 Two particles, *A* and *B*, are connected by a light inextensible string which passes over a smooth, fixed peg, as shown in the diagram.

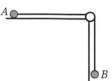

See p 45 Example 2 or review Advancing Maths for AQA M1 p 101

The particle *A*, of mass 0.6 kg, is in contact with a smooth horizontal surface, and the particle *B*, of mass 0.1 kg, hangs freely above the ground. The system is released from rest with the string taut and *A* moves towards the peg.

It can be assumed that, during the subsequent motion, *A* does **not** reach the peg.

While the particles move freely, the string is taut.

(a) Show that the acceleration of the particles is $1.4\,\mathrm{m\,s^{-2}}$.

(b) Find the tension in the string.

(c) Find the magnitude of the resultant force on the peg due to the tension in the string. [A]

Test yourself ANSWERS

5 **(b)** 0.84 N   **(c)** 1.19 N

4 **(a)** $6.31\,\mathrm{m\,s^{-2}}$ **(b)** 105 N

3 **(a)** 1570 N   **(b)** 210 N

2 **(a)** 35.3 N   **(b)** $2.24\,\mathrm{m\,s^{-2}}$ **(c)** 75.6 N   **(d)** $2.32\,\mathrm{m\,s^{-2}}$

1 **(a)** $2.33\,\mathrm{m\,s^{-2}}$ **(b)** 3.00 s   **(c)** 10.5 m

CHAPTER 6

Projectiles

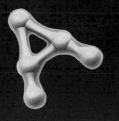

Key points to remember

1 The horizontal, x, and vertical, y, displacements of a projectile at time t are given by,

$$x = V \cos \theta t \quad \text{and} \quad y = V \sin \theta t - \frac{1}{2}gt^2$$

where V is the initial speed of the projectile and θ is the angle between the initial velocity and the horizontal.

2 The horizontal and vertical components of the velocity of a projectile are given by,

$$v_x = V \cos \theta \quad \text{and} \quad v_y = V \sin \theta - gt$$

3 The range is the horizontal distance travelled by the projectile.

4 At its maximum height the vertical component of the velocity of a projectile is zero.

6

Worked example 1

A ball is hit from ground level with an initial velocity of 20 m s^{-1} at an angle of $60°$ above the horizontal.

(a) Find the time of flight of the ball.

(b) Find the range of the ball.

(c) State two assumptions that you have made to obtain these values.

(a) The height of the ball at time t is given by,

$$y = 20 \sin 60°t - \frac{1}{2}gt^2$$

Using **1**

When the ball hits the ground, the height will be zero, which gives the equation below.

$$20 \sin 60°t - \frac{1}{2}gt^2 = 0$$

$$t\left(20 \sin 60° - \frac{1}{2}gt\right) = 0$$

$$t = 0 \quad \text{or} \quad 20 \sin 60° - \frac{1}{2}gt = 0$$

$$t = 0 \quad \text{or} \quad t = \frac{2 \times 20 \sin 60°}{g} = 3.5348 = 3.53 \text{ s (to 3 s.f.)}$$

The time of flight is 3.53 seconds.

(b) The horizontal displacement of the ball is given by,

$$x = 20 \cos 60° t$$

The range is found by substituting the time of flight into the expression for the horizontal displacement.

$$\text{Range} = 20 \cos 60° \times 3.5348$$
$$= 35.3 \, \text{m (to 3 s.f.)}$$

Using **1**

Using **3**

(c) It has been assumed that the ball is not subject to any forces other that its weight.
It has also been assumed that the ground is horizontal.

Worked example 2

A bullet is fired horizontally from a height of 5 metres, with a speed of 80 m s^{-1}. The ground is horizontal.

(a) Find the time for which the bullet is in the air.

(b) Find the horizontal distance travelled by the bullet.

(a) The height of the bullet is given by,

$$h = 5 - \frac{1}{2}gt^2$$

When it hits the ground the height will be zero, which enables t to be found.

$$5 - \frac{1}{2}gt^2 = 0$$

$$\frac{1}{2}gt^2 = 5$$

$$t = \sqrt{\frac{5 \times 2}{g}} = 1.0102 = 1.01 \, \text{s (to 3 s.f.)}$$

Using **1**

(b) The horizontal distance travelled is given by vt, which can be used with the time of flight to find the horizontal distance.

$$\text{Horizontal distance} = 80 \times 1.0102$$
$$= 80.8 \, \text{m (to 3 s.f.)}$$

Worked example 3

A golf ball is hit from a tee with a velocity of 45 m s⁻¹ at an
angle of 30° above the horizontal. It lands on a fairway, which is
3 metres lower than the tee.

(a) Find the time of flight of the ball.

(b) Find the range of the ball.

(c) Find the speed of the ball when it hits the ground.

(a) The height of the ball, relative to the level of the tee is
given by,

$$y = 45 \sin 30° t - \frac{1}{2}gt^2$$

Using **1**

When the ball hits the fairway, it is 3 metres lower than
the tee, so $y = -3$.

$$-3 = 45 \sin 30° t - \frac{1}{2}gt^2$$

$$\frac{1}{2}gt^2 - 45 \sin 30° t - 3 = 0$$

$$4.9t^2 - 45 \sin 30° t - 3 = 0$$

This quadratic equation needs to be solved to find the
time of flight.

$$t = \frac{45 \sin 30° \pm \sqrt{(45 \sin 30°)^2 - 4 \times 4.9 \times (-3)}}{2 \times 4.9}$$

$$= 4.7215 \text{ or } -0.1297$$

The time of flight will be 4.72 seconds to 3 s.f.

(b) The range of the ball is found by calculating the
horizontal displacement at the time that the ball lands.
This gives,

$$\text{Range} = 45 \cos 30° \times 4.7215 = 184 \text{ m (to 3 s.f.)}$$

Using **3**

(c) The velocity when the ball hits the ground has a
horizontal component. This is given by,

$$v_x = 45 \cos 30° = 38.97$$

Using **2**

The vertical component is found by using the time when
the ball hits the ground.

$$v_y = 45 \sin 30° - 9.8 \times 4.7215 = -23.77$$

Using **2**

The speed is given by the magnitude of the velocity,

$$v = \sqrt{38.97^2 + (-23.77)^2} = 45.6 \text{ m s}^{-1} \text{ (to 3 s.f.)}$$

6

Worked example 4

The unit vectors **i** and **j** are horizontal and vertical respectively. A particle is projected from ground level with velocity $(9\mathbf{i} + 6\mathbf{j})$ m s^{-1}. If the particle is initially at the origin, find an expression for the position vector of the particle at time t seconds. Use this to find the range of the projectile on horizontal ground.

Using $r = \mathbf{u}t + \dfrac{1}{2}\mathbf{a}t^2$, with $\mathbf{u} = 9\mathbf{i} + 6\mathbf{j}$ and $\mathbf{a} = -9.8\mathbf{j}$, gives,

$$\mathbf{r} = (9\mathbf{i} + 6\mathbf{j})t + \frac{1}{2}(-9.8\mathbf{j})t^2$$

$$= 9t\mathbf{i} + (6t - 4.9t^2)\mathbf{j}$$

The particle hits the ground when,

$$6t - 4.9t^2 = 0$$

$$t(6 - 4.9t) = 0$$

$$t = 0 \text{ or } t = \frac{6}{4.9} = 1.224$$

This value for t can now be used to calculate the range.

Using **3**

$$\text{Range} = 9 \times 1.224 = 11.0 \text{ m s}^{-1} \text{ (to 3 s.f.)}$$

Worked example 5

An arrow is fired with a velocity of 30 m s^{-1} at an angle of 60° above the horizontal from a height of 1.5 metres. Model the arrow as a particle that moves under gravity and find the maximum height of the arrow.

At its maximum height the vertical component of the velocity will be zero.

Using **4**

$$30 \sin 60° - 9.8t = 0$$

$$t = \frac{30 \sin 60°}{9.8}$$

This value can be substituted into the expression for the height of the arrow, to give,

$$y = 30 \sin 60° \times \left(\frac{30 \sin 60°}{9.8}\right) - 4.9\left(\frac{30 \sin 60°}{9.8}\right)^2$$

Using **1**

$$= 34.4 \text{ m (to 3 s.f.)}$$

Add the initial height to find the maximum height of the arrow.

$$\text{Maximum height} = 34.4 + 1.5 = 35.9 \text{ m}$$

REVISION EXERCISE 6

1 A ball is hit from ground level on a horizontal surface. The initial velocity of the ball is 12 m s^{-1} at an angle of 70° above the horizontal. Assume that the ball is a particle that moves freely under gravity.

   (a) Find the time of flight of the ball.

   (b) Find the range of the ball.

2 A bullet is fired with velocity 120 m s^{-1} at an angle of 25° above the horizontal, from ground level. Assume that the bullet is a particle that experiences no air resistance.

   (a) Find the time of flight of the bullet.

   (b) Find the range of the bullet.

3 A kangaroo leaves the ground travelling at 9 m s^{-1} and 8° above the horizontal. The ground is horizontal. Model the kangaroo as a particle that is not subject to air resistance.

   (a) Find the time that the kangaroo is in the air.

   (b) Find the length of the kangaroo's jump.

   (c) Find the maximum height of the kangaroo's jump.

4 A football is kicked from ground level with a velocity of 15 m s^{-1} at an angle of 50° above the horizontal. Assume that the ground is horizontal and that the ball is a particle that is not subject to air resistance.

   (a) Find the height of the ball when it has been moving for 0.8 seconds.

   (b) Find the horizontal distance travelled by the ball when it has been moving for 0.8 seconds.

   (c) Find the time it takes the ball to travel 9 metres horizontally.

   (d) Find the height of the ball when it has travelled 9 metres horizontally.

5 When a football is kicked from a free kick, it leaves the ground travelling at 26 m s^{-1} at an angle of 16° above the horizontal. The ball is 30 metres from the goal when it is kicked.

   (a) Find the time that it takes the ball to reach the goal.

   (b) Find the height of the ball when it crosses the goal line.

   (c) State two assumptions that you have made about the ball.

6 A tennis ball is hit from a height of 2.2 metres above ground level. The ball initially moves at 20 m s^{-1} at an angle of 5° above the horizontal. Find the maximum height of the ball above ground level.

6

7 An arrow is fired horizontally from a height of 1.8 metres with speed $17 \, \text{m s}^{-1}$. Model the arrow as a particle that moves under gravity.

  (a) Find the time it takes for the arrow to reach the ground.

  (b) Find the range of the arrow.

  (c) Find the speed of the arrow when it hits the ground.

8 The unit vectors **i** and **j** are horizontal and vertical respectively. A projectile is launched from ground level with velocity $(9\mathbf{i} + 15\mathbf{j}) \, \text{m s}^{-1}$.

  (a) Find the time of flight of the projectile.

  (b) Find the range of the projectile.

  (c) Find the maximum height of the projectile.

9 A stone is thrown horizontally from a quayside at a speed of $8 \, \text{m s}^{-1}$. Model the stone as a particle that moves under the influence of gravity only. The level of the water next to the quay is 6 metres below the initial position of the stone.

  (a) Find the time it takes the stone to reach the water.

  (b) Find the distance of the stone from the quayside when it hits the water.

  (c) Find the speed of the stone when it hits the water.

10 A shell is fired with a velocity of $90 \, \text{m s}^{-1}$ at an angle of $21°$ above the horizontal. Model the shell as a particle that experiences no air resistance.

  (a) Find the maximum height of the shell.

  (b) Find the speed of the shell when it is at its maximum height.

11 A bullet is fired with velocity $85 \, \text{m s}^{-1}$ at an angle of $20°$ above the horizontal. The bullet hits a target when it has travelled a horizontal distance of 100 metres. Model the bullet as a particle that moves freely under gravity.

  (a) Find the time that it takes the bullet to reach the target.

  (b) Find the difference between the height of the point where the bullet hits the target and the initial height of the bullet.

12 A shell is fired at a speed of $90 \, \text{m s}^{-1}$ at an angle of $40°$ above the horizontal. Model the shell as a particle that moves freely under gravity.
Find the range of the shell if it lands at:

  (a) the same height as it was fired;

  (b) a point 10 metres higher than the level at which it was fired.

13 A golf ball is hit so that its initial velocity is 20 m s^{-1} at an angle 36° above the horizontal. It lands at a point 3 metres higher than the point at which it was hit.

   (a) Find the time of flight of the ball.

   (b) Find the horizontal distance travelled by the ball.

14 A bullet is fired with velocity 80 m s^{-1} at an angle of 7° above the horizontal and from a height of 1.6 metres. Assume that the bullet moves freely under gravity and that the ground is horizontal.

   (a) Find the time of flight of the bullet.

   (b) Find the range of the bullet.

15 A basketball is thrown with velocity 6 m s^{-1} at an angle of 45° above the horizontal and from a height of 1.2 metres above ground level. It lands on the ground.

   (a) Find the time of flight of the ball.

   (b) Find the horizontal distance travelled by the ball.

16 The unit vectors **i** and **j** are horizontal and vertical respectively. A projectile is launched from ground level with velocity $(20\mathbf{i} + 12\mathbf{j})$ m s^{-1}. Find the maximum height of the projectile.

17 The unit vectors **i** and **j** are horizontal and vertical respectively. A projectile is launched from ground level with velocity $(U\mathbf{i} + V\mathbf{j})$ m s^{-1}. It hits the ground having travelled 20 metres horizontally in 3 seconds. Find U and V.

18 A ball is kicked from ground level with velocity 16 m s^{-1} at an angle of 40° above the horizontal. It is headed by another player when it is at a height of 2 metres.

   (a) Find the two possible times between the ball being kicked and headed.

   (b) Find the horizontal distance travelled by the ball in each of these times.

19 A bullet is fired with velocity 85 m s^{-1} at an angle of 70° above the horizontal. Assume that the bullet moves freely under gravity and that it is fired from ground level. Find the maximum height of the bullet.

20 The unit vectors **i** and **j** are horizontal and vertical respectively. A projectile is launched from the point with position vector $(5\mathbf{j})$ metres with velocity $(8\mathbf{i} + 10\mathbf{j})$ m s^{-1}. The origin is on a horizontal surface.

   (a) Find the time of flight of the projectile.

   (b) Find the range of the projectile.

21 A bullet is fired at 100 m s^{-1}, and at an angle of $5°$ below the horizontal, from a cliff top which is at a height of 20 metres above sea level.

(a) Find the time that it takes for the bullet to reach the sea.

(b) Find the horizontal distance travelled by the bullet.

(c) Find the speed of the bullet when it hits the sea.

22 A particle is projected horizontally at a height of 5 metres above ground level. Assume that the particle moves freely under gravity.

(a) Find the time that it takes the particle to reach the ground.

(b) If the particle travels a horizontal distance of 24 metres, find the initial velocity of the particle.

(c) Find the velocity of the particle when it hits the ground.

23 A particle is projected with velocity 12 m s^{-1} at an angle α above the horizontal. The particle starts at ground level and lands at the same level 0.9 seconds later, having travelled 7 metres horizontally. Find α.

24 A projectile has initial velocity $V \text{ m s}^{-1}$ at an angle $40°$ above the horizontal. It travels a horizontal distance of 40 metres in 1.8 seconds. Find V.

25 The unit vectors **i** and **j** are horizontal and vertical respectively. A projectile is launched with velocity $(U\mathbf{i} + V\mathbf{j}) \text{ m s}^{-1}$.

(a) Find the time of flight of the projectile in terms of V and g.

(b) Find the range of the projectile in terms of U, V and g.

26 The unit vectors **i** and **j** are horizontal and vertical respectively. A projectile is launched with velocity $(U\mathbf{i} + 2U\mathbf{j}) \text{ m s}^{-1}$. Find the maximum height of the projectile in terms of U and g.

27 A football is kicked from horizontal ground. It initially moves with speed 20 m s^{-1} at an angle of $30°$ above the horizontal.

(a) (i) Show that the ball hits the ground approximately 2.04 seconds after it has been kicked.

(ii) Hence find the range of the ball.

(b) In fact the ball comes into contact with a player's head when it is at a height of 2 metres. Find the speed of the ball at this height. [A]

28 A football is placed on a horizontal surface and kicked, so that it has an initial velocity of 12 m s^{-1} at an angle of 40° above the horizontal.

(a) State two modelling assumptions that it would be appropriate to make when considering the motion of the football.

(b) (i) Find the time that it takes for the ball to reach its maximum height.
   (ii) Hence, show that the maximum height of the ball is 3.04 metres correct to three significant figures.

(c) After the ball has reached its maximum height it hits the bar of a goal at a height of 2.44 metres. Find the horizontal distance of the goal from the point where the ball was kicked.   [A]

29 A golf ball is struck so that its initial velocity is 30 m s^{-1} at an angle of 60° above the horizontal. Assume that the ground is horizontal.

(a) (i) Show that the time of flight of the particle is 5.30 seconds, correct to three significant figures.
   (ii) Find the distance of the ball from its initial position when it hits the ground for the first time.

(b) In fact the ball hits a tree at a point, which is at a height of 5 metres above the ground. Given that the ball is descending when it hits the tree, calculate the distance of the tree from the point where the ball was struck.   [A]

30 A golf ball is struck at a point O on a horizontal stretch of ground and moves with an initial velocity of 15 m s^{-1} at an angle of θ to the horizontal, where $\cos \theta = 0.8$. The ball subsequently lands at a point M which is at a higher horizontal level than O.

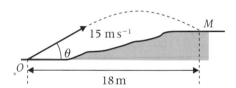

The horizontal distance between the points O and M is 18 metres, as shown in the diagram.

(a) Find the time the ball takes to travel from O to M.

(b) (i) Find the vertical component of the velocity of the ball when it reaches M.
   (ii) Find the angle the direction of motion of the ball makes with the horizontal when it reaches M.   [M]

Test yourself	**What to review**
	If your answer is incorrect:
1 A golf ball is hit from ground level with a velocity of 32 m s^{-1} at 24° above the horizontal. Model the golf ball as a particle that moves freely under gravity. Also assume that the ground is horizontal. **(a)** Find the time of flight of the ball. **(b)** Find the range of the ball.	See p 65 Example 1 or review Advancing Maths for AQA M1 pp 112–113
2 A stuntman drives a car off the top of a vertical cliff, which is 4 metres above sea level. Assume that the car moves freely under gravity and that it was travelling horizontally at 20 m s^{-1} when it left the cliff. **(a)** Find the time that it takes for the car to reach the sea. **(b)** Find the distance of the car from the cliff when it hits the sea.	See p 66 Example 3 or review Advancing Maths for AQA M1 p 119
3 A stone is fired from a catapult, at a speed of 24 m s^{-1} at an angle of 30° above the horizontal. When it has travelled 20 metres horizontally the stone hits a vertical wall. **(a)** Find the time that it takes the stone to reach the wall. **(b)** Find the height of the stone when it hits the wall, assuming that it is initially at ground level. **(c)** Find the speed of the stone when it hits the wall.	See p 66 Examples 2 and 3 or review Advancing Maths for AQA M1 pp 116–118
4 A cricket ball is hit from ground level with velocity 12 m s^{-1} at an angle of 70° above the horizontal. As the ball is descending it is caught at a height of 1.4 metres. **(a)** Find the horizontal distance the ball has travelled when it is caught. **(b)** Find the speed of the ball when it is caught.	See p 66 Examples 2 and 3 or review Advancing Maths for AQA M1 pp 116–118
5 A golf ball is hit so that it initially travels at 32 m s^{-1} at an angle of 60° above the horizontal. The ball lands on a horizontal surface 5 metres higher than the ground from which it was hit, as shown in the diagram.	See p 66 Example 3 or review review Advancing Maths for AQA M1 pp 116–118

(a) Show that the ball is in the air for approximately 5.47 seconds.

(b) Calculate the horizontal distance travelled by the ball.

(c) Find the speed of the ball when it hits the ground.   [A]

Test yourself ANSWERS

1 (a) 2.66 s   **(b)** 77.7 m

2 (a) 0.904 s   **(b)** 18.1 m

3 (a) 0.962 s   **(b)** 7.01 m   **(c)** 20.9 m s^{-1}

4 (a) 8.90 m   **(b)** 10.8 m s^{-1}

5 (b) 87.5 m   **(c)** 30.4 m s^{-1}

6

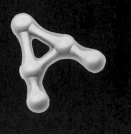

CHAPTER 7
Conservation of momentum

Key points to remember

1   Momentum is conserved in collisions.

2   Use the equation $m_A u_A + m_B u_B = m_A v_A + m_B v_B$ for collisions on a straight line.

3   Use $m_A \mathbf{u}_A + m_B \mathbf{u}_B = m_A \mathbf{v}_A + m_B \mathbf{v}_B$ when working with vectors in two dimensions.

Worked example 1

Two particles A and B are moving on a straight line. The particle A has mass 6 kg and velocity 5 m s^{-1} and B has mass 4 kg and velocity 3 m s^{-1}. They collide and coalesce. Find the velocity of the particles after the collision.

The diagram shows the velocities of the particles before and after the collision.

$$6\,\text{kg} \qquad 4\,\text{kg} \qquad\qquad\qquad 10\,\text{kg}$$

A ------- B - - - - - - - - - - - ○ - - - -

$$5\,\text{m s}^{-1} \quad\; 3\,\text{m s}^{-1} \qquad\qquad\quad v\,\text{m s}^{-1}$$

BEFORE   AFTER

Conservation of momentum gives,

$$6 \times 5 + 4 \times 3 = (6 + 4)v$$

$$42 = 10v$$

$$v = \frac{42}{10} = 4.2 \text{ m s}^{-1}$$

Using **1**

Using **2**

Worked example 2

A car, of mass 1000 kg, is travelling at 8 m s^{-1}, when it collides with a lorry of mass 3000 kg travelling in the opposite direction at 5 m s^{-1}. After the collision the two vehicles move together. Find the velocity of the vehicles after the collision.

Modelling the vehicles as particles, gives the diagram below.

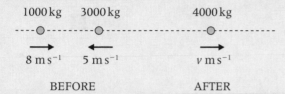

Using conservation of momentum gives,

$$1000 \times 8 + 3000 \times (-5) = (1000 + 3000)v$$

$$8000 - 15\,000 = 4000v$$

$$-7000 = 4000v$$

$$v = \frac{-7000}{4000} = -1.75 \text{ m s}^{-1}$$

Using **1**

Using **2**

Note the use of negative signs for velocities that are directed towards the left-hand side of the diagram. In this collision the direction of the velocity of the car is reversed.

7

Worked example 3

The diagram shows the velocities of two particles, A and B, before and after a collision. The mass of particle B is m kg.

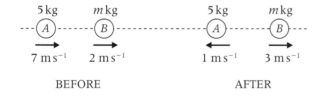

Find m.

Using conservation of momentum gives,

$$5 \times 7 + m \times 2 = 5 \times (-1) + m \times 3$$

$$35 + 2m = -5 + 3m$$

$$m = 40$$

Using **1**

Using **2**

Worked example 4

Two particles, A and B, are moving with velocities $(8\mathbf{i} + 2\mathbf{j})$ m s^{-1} and $(4\mathbf{i} - 6\mathbf{j})$ m s^{-1} respectively. The mass of A is 7 kg and the mass of B is 9 kg. The particles collide and coalesce. Find the velocity of the particles after the collision.

Let $\mathbf{v}$ be the velocity after the collision. Using conservation of momentum gives,

$$7(8\mathbf{i} + 2\mathbf{j}) + 9(4\mathbf{i} - 6\mathbf{j}) = (7 + 9)\mathbf{v}$$

$$56\mathbf{i} + 14\mathbf{j} + 36\mathbf{i} - 54\mathbf{j} = 16\mathbf{v}$$

$$92\mathbf{i} - 40\mathbf{j} = 16\mathbf{v}$$

$$\mathbf{v} = \frac{92\mathbf{i} - 40\mathbf{j}}{16} = 5.75\mathbf{i} - 2.5\mathbf{j}$$

Using **1**

Using **3**

Worked example 5

A particle has velocity $(9\mathbf{i} - 2\mathbf{j})$ m s^{-1} and mass 4 kg. It collides with a second particle which has mass m kg and velocity $(-6\mathbf{i} + 4\mathbf{j})$ m s^{-1}. The particles coalesce during the collision and after the collision move parallel to the unit vector $\mathbf{j}$.

(a) Find m.

(b) Find the velocity after the collision.

After the collision the velocity will be $k\mathbf{j}$ m s^{-1}, where k is a constant. Using conservation of momentum gives,

$$4(9\mathbf{i} - 2\mathbf{j}) + m(-6\mathbf{i} + 4\mathbf{j}) = (4 + m) \times k\mathbf{j}$$

$$36\mathbf{i} - 8\mathbf{j} - 6m\mathbf{i} + 4m\mathbf{j} = (4 + m) \times k\mathbf{j}$$

$$(36 - 6m)\mathbf{i} + (4m - 8)\mathbf{j} = (4 + m) \times k\mathbf{j}$$

(a) As the $\mathbf{i}$ component of the velocity after the collision is zero,

$$36 - 6m = 0$$

$$6m = 36$$

$$m = 6$$

Using **1**

Using **3**

(b) Now considering the **j** component and substituting $m = 6$ gives,

$$4m - 8 = (4 + m)k$$

$$4 \times 6 - 8 = (4 + 6)k$$

$$16 = 10k$$

$$k = \frac{16}{10} = 1.6$$

The velocity after the collision is $1.6\mathbf{j}\text{ m s}^{-1}$.

Worked example 6

Two particles, A and B, which have masses of 4 kg and 7 kg respectively collide. Before the collision the velocity of B was $\begin{bmatrix} 6 \\ -7 \end{bmatrix}\text{ m s}^{-1}$. After the collision the velocity of A is $\begin{bmatrix} 2 \\ 1 \end{bmatrix}\text{ m s}^{-1}$ and the velocity of B is $\begin{bmatrix} 3 \\ -2 \end{bmatrix}\text{ m s}^{-1}$. The velocity of A before the collision was $\mathbf{v}\text{ m s}^{-1}$. Find $\mathbf{v}$.

Using conservation of momentum gives,

$$4\mathbf{v} + 7\begin{bmatrix} 6 \\ -7 \end{bmatrix} = 4\begin{bmatrix} 2 \\ 1 \end{bmatrix} + 7\begin{bmatrix} 3 \\ -2 \end{bmatrix}$$

$$4\mathbf{v} + \begin{bmatrix} 42 \\ -49 \end{bmatrix} = \begin{bmatrix} 8 \\ 4 \end{bmatrix} + \begin{bmatrix} 21 \\ -14 \end{bmatrix}$$

$$4\mathbf{v} + \begin{bmatrix} 42 \\ -49 \end{bmatrix} = \begin{bmatrix} 29 \\ -10 \end{bmatrix}$$

$$4\mathbf{v} = \begin{bmatrix} -13 \\ 39 \end{bmatrix}$$

$$\mathbf{v} = \frac{1}{4}\begin{bmatrix} -13 \\ 39 \end{bmatrix} = \begin{bmatrix} -3.25 \\ 9.75 \end{bmatrix}$$

Using **1**

Using **3**

7

REVISION EXERCISE 7

1 Two particles A and B are moving on a straight line when they collide. The particle A has mass 7 kg and velocity 2 m s^{-1} and B has mass 3 kg and velocity 1 m s^{-1}. During the collision the particles coalesce. Find the velocity of the particles after the collision.

2 The diagram shows the velocities of two particles, *A* and *B*, before and after a collision. The velocity of particle *B* after the collision is *v* m s⁻¹.

Find *v*.

3 Two particles have masses of 10 kg and 6 kg respectively and are moving towards each other along a straight line, both with speed 5 m s⁻¹. During the collision the 6 kg mass changes direction and its speed is reduced to 3 m s⁻¹. Describe how the 10 kg particle moves after the collision.

4 Two particles of equal mass have speeds of 2 m s⁻¹ and 4 m s⁻¹. They move along a straight line and collide with each other. After the collision they move together. Find the speed after the collision if they are:

(a) moving in the same direction;

(b) moving in opposite directions.

5 A van, of mass 1600 kg, collides with a stationary car of mass 1000 kg. After the collision the two vehicles move together at 4 m s⁻¹. Find the speed of the van before the collision.

6 Two train trucks are on a length of track. One is initially stationary and the other is moving at 8 m s⁻¹. The stationary truck has a mass of 40 tonnes. The moving truck collides with the stationary truck and they couple together and move at 2.5 m s⁻¹. Find the mass of the truck that was initially moving.

7 Two particles, *A* and *B*, are moving directly towards each other with speeds of 3 m s⁻¹ and 2 m s⁻¹ respectively. The mass of *A* is 7 kg and the mass of *B* is 4 kg. After the collision the speed of *A* is 1 m s⁻¹. Find the two possible speeds of *B* after the collision.

8 A boy, of mass 40 kg, stands on a skateboard of mass 2 kg. The boy jumps off the skateboard with a horizontal speed of 1.2 m s⁻¹. Calculate the speed at which the skateboard moves if momentum is conserved.

9 A bullet is fired horizontally from a gun at a speed of 90 m s⁻¹. The mass of the bullet is 80 grams and the mass of the gun is 1.2 kg. Find the speed at which the gun recoils.

10 A child stands on a trolley and catches a heavy ball that is travelling horizontally. The combined mass of the child and the trolley is 40 kg and the ball is travelling at 6 m s^{-1} when it is caught. After the collision the child, ball and trolley all move at 1 m s^{-1}. Find the mass of the ball.

11 Three smooth spheres, A, B and C, lie on a straight, horizontal line. They have masses of 3 kg, 2 kg and 0.5 kg respectively. Sphere A is set into motion with a velocity of 2 m s^{-1}. When it collides with B its velocity is halved. When B collides with C its velocity is halved.

    (a) Find the velocity of B after it has been hit by A.

    (b) Find the velocity of C after it has been hit by B.

    (c) State with a reason, whether or not any more collisions take place.

12 Three particles are initially at rest in a line on a smooth horizontal surface. The first particle, which has mass 4 kg, is set into motion with a velocity of 8 m s^{-1} towards the second particle, which has mass 2 kg. The two particles collide and the velocity of the first particle is reduced to 3 m s^{-1}. The second particle then collides with the third particle, which has mass 1 kg. During this collision the two particles coalesce.

    (a) Find the velocities of the second particle after the first collision.

    (b) Determine whether or not there is a third collision.

7

13 Two particles, A and B, are moving with velocities $(4\mathbf{i} + 12\mathbf{j}) \text{ m s}^{-1}$ and $(9\mathbf{i} - \mathbf{j}) \text{ m s}^{-1}$ respectively. The mass of A is 20 kg and the mass of B is 30 kg. The particles collide and coalesce. Find the velocity of the particles after the collision.

14 Two particles, A and B, are moving with velocities $\mathbf{v} \text{ m s}^{-1}$ and $(4\mathbf{i} - 6\mathbf{j}) \text{ m s}^{-1}$ respectively. The mass of A is twice the mass of B. The particles collide and coalesce during the collision. The velocity of the particles after the collision is $(2\mathbf{i} + \mathbf{j}) \text{ m s}^{-1}$. Find $\mathbf{v}$.

15 Two particles, A and B, which have masses of 12 kg and 8 kg respectively collide. Before the collision the velocity of A was $\begin{bmatrix} 7 \\ -2 \end{bmatrix} \text{ m s}^{-1}$ and the velocity of B was $\begin{bmatrix} 5 \\ 6 \end{bmatrix} \text{ m s}^{-1}$. After the collision the particles move together with velocity $\mathbf{v} \text{ m s}^{-1}$. Find $\mathbf{v}$.

16 Two particles, A and B, which have masses of 5 kg and 15 kg respectively collide. Before the collision the velocity of B was $\begin{bmatrix} -4 \\ -1 \end{bmatrix}$ m s^{-1}. After the collision the velocity of A is $\begin{bmatrix} 2 \\ 0 \end{bmatrix}$ m s^{-1} and the velocity of B is $\begin{bmatrix} 3 \\ -1 \end{bmatrix}$ m s^{-1}. The velocity of A before the collision was $\mathbf{v}$ m s^{-1}. Find $\mathbf{v}$.

17 A particle with mass 3 kg and velocity $(4\mathbf{i} + 7\mathbf{j})$ m s^{-1} collides with a stationary particle of mass 5 kg. After the collision the 3 kg particle has velocity $2\mathbf{j}$ m s^{-1}. Find the velocity of the other particle after the collision.

18 As the result of a collision with a stationary particle of mass 4 kg, the velocity of a particle with mass 2 kg changes from $(9\mathbf{i} + 12\mathbf{j})$ m s^{-1} to $(\mathbf{i} + 2\mathbf{j})$ m s^{-1}. Find the velocity of the stationary particle after the collision.

19 A particle, of mass 3 kg, is moving with velocity $(2\mathbf{i} - 2\mathbf{j})$ m s^{-1} when it collides with a particle, of mass 7 kg, moving with velocity $V(\mathbf{i} + 3\mathbf{j})$ m s^{-1}. After the collision, the particles move together in a direction parallel to the unit vector $\mathbf{i}$.

  (a) Find V giving your answer as a fraction.

  (b) Find the speed of the particles after the collision.

20 Two particles, A and B, are moving with velocities $(8\mathbf{i} + 2\mathbf{j})$ m s^{-1} and $(4\mathbf{i} - 6\mathbf{j})$ m s^{-1} respectively. The mass of A is 4 kg and the mass of B is 9 kg. The particles collide and after the collision the velocity of B is $(6\mathbf{i} - 4\mathbf{j})$ m s^{-1}. Find the velocity of particle A after the collision.

21 Two particles have masses of 5 kg and 2 kg and velocities $\begin{bmatrix} X \\ 0 \end{bmatrix}$ m s^{-1} and $\begin{bmatrix} 0 \\ Y \end{bmatrix}$ m s^{-1} respectively. The particles collide. After the collision the particles move together with velocity $\begin{bmatrix} 7 \\ 4 \end{bmatrix}$ m s^{-1}. Find X and Y.

22 Two particles, A and B, are moving with velocities $(4\mathbf{i} + 8\mathbf{j})$ m s^{-1} and $(2\mathbf{i} - 5\mathbf{j})$ m s^{-1} respectively. The mass of A is 2 kg and the mass of B is 8 kg. The particles collide and after the collision the velocity of A is $(2\mathbf{i} + a\mathbf{j})$ m s^{-1} and the velocity of B is $(b\mathbf{i} - 2\mathbf{j})$ m s^{-1}. Find a and b.

23 A particle, of mass 11 kg, has velocity $(7\mathbf{i} - 3\mathbf{j})$ m s^{-1}. It collides with a second particle which has mass m kg and velocity $(-2\mathbf{i} + 3\mathbf{j})$ m s^{-1}. The particles coalesce during the collision and after the collision move parallel to the unit vector $\mathbf{i}$.

    (a) Find m.

    (b) Find the speed of the particles after the collision.

24 Two particles have masses of m and M kg and velocities $\begin{bmatrix} 5 \\ 1 \end{bmatrix}$ m s^{-1} and $\begin{bmatrix} 2 \\ k \end{bmatrix}$ m s^{-1} respectively. The particles collide.

After the collision the particles move together with velocity . $\begin{bmatrix} 3.2 \\ 2 \end{bmatrix}$ m s^{-1} Find k, giving your answer as a fraction.

25 A car is travelling due north at a speed of 20 m s^{-1} and has mass 1200 kg. A van is travelling due east at a speed of 12 m s^{-1} and has mass 2800 kg. The unit vectors $\mathbf{i}$ and $\mathbf{j}$ are directed east and north respectively.

    (a) Write down the velocity of the car.

    (b) Write down the velocity of the van.

    (c) The car and van collide and move together after the collision. Find their speed after the collision.

26 Two particles are travelling along perpendicular paths when they collide. One has mass 4 kg and speed 30 m s^{-1}, while the other has mass 16 kg and speed 5 m s^{-1}. During the collision the particles are deflected through 90° angles. Find the speeds of the two particles after the collision.

27 A particle, A, of mass 12 kg is moving on a smooth horizontal surface with velocity $\begin{bmatrix} 4 \\ 7 \end{bmatrix}$ m s^{-1}, when it collides and coalesces with a second particle, B, of mass 4 kg.

    (a) If before the collision the velocity of B was $\begin{bmatrix} 2 \\ 3 \end{bmatrix}$ m s^{-1}, find the velocity of the combined particle after the collision.

    (b) If after the collision the velocity of the combined particle is $\begin{bmatrix} 1 \\ 4 \end{bmatrix}$ m s^{-1}, find the velocity of B before the collision.

                                                                 collision. [A]

7

28 A child of mass 25 kg is sitting in a trolley of mass 13 kg. The trolley is initially at rest on a horizontal surface. A ball of mass 2 kg is thrown towards the child, who catches it. The ball is travelling horizontally at 5 m s⁻¹ just before it is caught. Assume that there is no resistance to the motion of the trolley.

  (a) Find the speed of the child and the trolley after the ball has been caught.

  (b) The child places the ball in the trolley. A second, identical ball is thrown, in the same direction as the first ball, and the child catches it. This ball is travelling horizontally at 6 m s⁻¹ just before it is caught. Find the speed of the child and the trolley after the second ball has been caught. [A]

Test yourself	What to review
	If your answer is incorrect:
1 A van, of mass 2500 kg, is travelling at 8 m s⁻¹, when it collides with a stationary car, of mass 1500 kg. After the collision they move together. Find their velocity immediately after the collision.	See p 56 Examples 1 and 2 or review Advancing Maths for AQA M1 p 128
2 Two particles A and B are moving in the same direction along a straight line. The particle A has mass m kg and velocity 10 m s⁻¹ and particle B has mass 4 kg and velocity 8 m s⁻¹. They collide and coalesce. After the collision the velocity of the particles is 8.4 m s⁻¹. Find m.	See p 58 Example 5 or review Advancing Maths for AQA M1 pp 128–129
3 Two particles, A and B, are moving with velocities $\begin{bmatrix} 3 \\ 4 \end{bmatrix}$ m s⁻¹ and $\begin{bmatrix} -2 \\ -5 \end{bmatrix}$ m s⁻¹ respectively. The mass of A is 17 kg and the mass of B is 23 kg. The particles collide and coalesce. Find the velocity of the particles after the collision.	See p 59 Example 6 or review Advancing Maths for AQA M1 pp 134–135
4 Two particles have masses of m and M kg and velocities $5\mathbf{i}$ m s⁻¹ and $3\mathbf{j}$ m s⁻¹ respectively. The particles collide. After the collision the particles move together with velocity $(2\mathbf{i} + k\mathbf{j})$ m s⁻¹. **(a)** Show that $m = \dfrac{2M}{3}$. **(b)** Find k, giving your answer as a fraction.	See p 57 Example 4 or review Advancing Maths for AQA M1 pp 135–137

Test yourself (continued)

If your answer is incorrect:

5 Three particles, A, B and C, are initially at rest in a straight line on a smooth horizontal surface. The masses of the particles are 2 kg, 4 kg and m kg respectively.

See p 57 Example 3 or review
Advancing Maths for AQA M1
pp 128–129

(a) Particle A is set in motion with speed 4 m s^{-1} directly towards particle B. After colliding with B, particle A continues to move in the same direction but with speed 1 m s^{-1}. Find the speed of B after this collision.

(b) Particle B then collides with particle C. After this collision, C moves with a speed of 2 m s^{-1}. Find the speed of B after this collision, giving your answer in terms of m.

(c) If A collides with B again, show that $m > 1$. [A]

Test yourself   ANSWERS

1 5 m s^{-1}

2 $m = 1$

3 $\begin{bmatrix} 0.125 \\ -1.175 \end{bmatrix} \text{ m s}^{-1}$

4 **(b)** $k = \dfrac{6}{5}$

5 **(a)** 1.5 m s^{-1} **(b)** $\dfrac{6 - 2m}{4}$

7

Examination style papers

Answer **all** questions.
Time allowed: 1 hour 30 minutes

PRACTICE PAPER I

1 A ball falls from rest at a height of 12 metres above ground level.

   (a) Find the speed of the ball when it hits the ground.

           (3 marks)

   (b) Find the time that it takes the ball to reach the ground.

           (2 marks)

   (c) Find the height of the ball above ground level when its speed is 5 m s^{-1}. (3 marks)

2 Two particles A and B have masses of 3 kg and 5 kg respectively. They are moving on a smooth horizontal surface when they collide. The velocity of A before the collision is $\begin{bmatrix} 2 \\ -2 \end{bmatrix} \text{ m s}^{-1}$ and the velocity of B before the collision is $\begin{bmatrix} -4 \\ 6 \end{bmatrix} \text{ m s}^{-1}$.

   (a) If the particles coalesce during the collision, find the velocity of the combined particles after the collision.

           (3 marks)

   (b) If after the collision the velocity of A is $\begin{bmatrix} 1 \\ 2 \end{bmatrix} \text{ m s}^{-1}$, find the velocity of B after the collision (3 marks)

3 In a stretch of open water a current flows west at 1 m s^{-1}. The velocity of a boat relative to the water is 3 m s^{-1} on a bearing of 325°.

   (a) Find the magnitude of the resultant velocity of the boat.

           (3 marks)

   (b) Find the bearing of the resultant velocity. (3 marks)

4 A particle, of mass 10 kg, is at rest on a rough slope inclined at 36° to the horizontal.

   (a) Find the normal reaction force acting on the particle.

           (2 marks)

   (b) Find the minimum value of the coefficient of friction between the particle and the slope. (4 marks)

5 A wooden block, of mass 8 kg, is at rest on a rough, horizontal table. A light string is attached to the block. The string passes over a smooth peg and the other end is attached to a particle of mass 5 kg. The coefficient of friction between the block and the table is 0.5. The force of magnitude P Newtons is applied as shown in the diagram to keep the system in equilibrium, with the block on the point of sliding towards the pulley.

(a) Find the magnitude of the friction force when the block is on the point of sliding. (2 marks)

(b) Find P. (3 marks)

(c) The force P is removed and the system begins to accelerate.
(i) Find the acceleration of the system.
(ii) Find the tension in the string. (7 marks)

6 A golf ball is struck so that it leaves the ground with a velocity of 42 m s^{-1} at $56°$ above the horizontal. The ball lands in a bunker at a point 3 metres lower than its starting point.

(a) Find the maximum height of the ball above ground level during its flight. (4 marks)

(b) Find the time that it takes for the golf ball to reach the bunker. (5 marks)

(c) Find the horizontal distance travelled by the golf ball. (2 marks)

7 The forces $\mathbf{F}_1 = 4\mathbf{i} + 8\mathbf{j}$, $\mathbf{F}_2 = -6\mathbf{i} + 2\mathbf{j}$ and $\mathbf{F}_3 = 3\mathbf{i} - 14\mathbf{j}$ act on a particle that is on a smooth horizontal surface. The vectors $\mathbf{i}$ and $\mathbf{j}$ are perpendicular and lie on the horizontal surface.

(a) Find the resultant of $\mathbf{F}_1$, $\mathbf{F}_2$ and $\mathbf{F}_3$. (2 marks)

(b) The mass of the particle is 5 kg. Show that the acceleration of the particle is $(0.2\mathbf{i} - 0.8\mathbf{j}) \text{ m s}^{-2}$. (2 marks)

(c) At time $t = 0$, the particle has velocity $5\mathbf{j}$ and the particle is at the origin. Find the position vector of the particle relative to the origin when it is moving parallel to the vector $\mathbf{i} + \mathbf{j}$. (6 marks)

Answer **all** questions.
Time allowed: 1 hour 30 minutes

PRACTICE PAPER 2

1 A ball is thrown vertically upwards from a height of 1.5 metres above ground level. Its initial velocity is $10 \, \text{m s}^{-1}$. Assume that the ball is a particle and that no resistance forces act on the ball as its moves.

    (a) Find the time that it takes for the ball to reach its maximum height.   (3 marks)

    (b) Find the maximum height of the ball.   (3 marks)

    (c) Find the speed of the ball when it hits the ground.   (3 marks)

2 The diagram shows three forces that act in a plane. The forces are in equilibrium.

    (a) Show that $P = 23.1$ Newtons, correct to three significant figures.   (3 marks)

    (b) Find Q.   (3 marks)

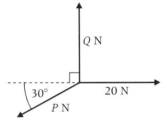

3 An aeroplane has a velocity relative to the air of $200 \, \text{m s}^{-1}$ on a bearing of 145°. The air is moving due east at $50 \, \text{m s}^{-1}$.

    (a) Find the magnitude of the resultant velocity of the aeroplane.   (4 marks)

    (b) Find the bearing of the resultant velocity.   (4 marks)

4 A brick, of mass 1.2 kg, is placed on a rough horizontal surface. A horizontal force of magnitude P Newtons is applied to the brick. The coefficient of friction between the brick and the surface is 0.5.

    (a) Find the magnitude of the friction force acting on the block if

        (i) $P = 4.7 \, \text{N}$

        (ii) $P = 7 \, \text{N}$ (3 marks)

    (b) Find the acceleration of the block in each of the above cases.   (4 marks)

5 Two particles have masses of 6 kg and 4 kg. They are connected by a light string that passes over a smooth fixed peg. They are released from rest from the positions shown in the diagram.

    (a) Find the acceleration of the particles.   (5 marks)

    (b) Show that the tension in the string is 47.0 N, correct to three significant figures.   (2 marks)

    (c) State one important assumption that you have made about the particles in order to obtain your answers.   (1 mark)

6 A child of mass 48 kg, is on a slide that is inclined at 40° to the horizontal. He accelerates at 0.5 m s^{-2}. Model the child as a particle and assume that there is no air resistance.

    (a) Draw a diagram to show the forces acting on the child.

                                                              (1 mark)

    (b) Find the magnitude of the normal reaction force acting on the child. (2 marks)

    (c) Find the magnitude of the friction reaction force acting on the child. (4 marks)

    (d) Find the coefficient of friction between the child and the slide. (2 marks)

    (e) Air resistance does in reality affect the motion of the child. Explain how this would change your answer to part **(d)** (2 marks)

7 A bullet is fired horizontally from a rifle at 80 m s^{-1}. The bullet is at a height of 3 metres above ground level when it is fired. Model the bullet as a particle that does not experience any air resistance.

    (a) Show that the bullet hits the ground when it has been travelling for 0.782 s, correct to three significant figures. (3 marks)

    (b) Find the horizontal distance travelled by the bullet. (2 marks)

    (c) Find the angle between the velocity of the bullet and the horizontal when its hits the ground. (4 marks)

    (d) Describe what happens to the horizontal component of the velocity of the bullet as it is moving. (1 mark)

    (e) Describe what happens to the magnitude of the vertical component of the velocity of the bullet as it is moving. (1 mark)

8 A particle moves with constant acceleration, so that at time t seconds its position vector relative to an origin O is $\mathbf{r}$ metres and its velocity is $\mathbf{v}$ m s^{-1}. When $t = 0$, $\mathbf{r} = 6\mathbf{i} + 2\mathbf{j}$ and $\mathbf{v} = -2\mathbf{i} + 3\mathbf{j}$. When $t = 4$, $\mathbf{r} = -0.4\mathbf{i} + 10.8\mathbf{j}$. The unit vectors $\mathbf{i}$ and $\mathbf{j}$ are directed east and north respectively.

    (a) Show that $a = (0.2\mathbf{i} - 0.4\mathbf{j})$ m s^{-2}. (4 marks)

    (b) Find an expression for the velocity of the particle at time t. (2 marks)

    (c) Find the time when the particle is travelling due south (3 marks)

    (d) Find the time when the particle is travelling north east, giving your answer as a fraction in its simplest form. (5 marks)

Answers

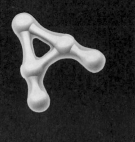

Where appropriate answers are given to 3 s.f.

1 (a) 3250 m   (b) 16.3 m s^{-1} (c) -0.313 m s^{-2}
2 (a) 187 m   (b) 4.66 m s^{-1} (c) 0.625 m s^{-2}
3 (a) 205 m   (b) 2.05 m s^{-1} (c) 145 m   (d) 1.45 m s^{-1}
4 (a) 152 m   (b) 7.6 m s^{-1} (c) 8 m   (d) -0.4 m s^{-1}
5 (a) 169.5 m   (b) 8.48 m s^{-1} (c) 1.28 m s^{-1}
6 (a) 20 m   (b) 8 m s^{-1} (c) 1.6 m s^{-2}
7 (a) 0.5 m s^{-2} (b) 10 m s^{-1}
8 (a) 0.2 m s^{-2} (b) 120 m
9 (a) -1.3 m s^{-2} (b) 98.5 m   (c) 12.3 s   (d) 60 m
10 (a) 20 s   (b) 19.2 m   (c) 0.96 m s^{-1}
11 (a) -5 m s^{-1} (b) 6 s   (c) 5 m s^{-1} (d) 50 m
12 2.4 s and 8.4 m
13 (a) 40 m   (b) 5.71 s
14 (a) 0.510 s   (b) 9.28 m   (c) 1.89 s   (d) 13.5 m s^{-1}
15 (a) 1.11 s   (b) 10.8 m s^{-1}
16 (a) 8.82 m s^{-1} (b) 3.97 m
17 (a) 7.35 m   (b) 1.22 s
18 (a) 22.5 m   (b) 4.29 s   (c) 3.78 s
19 (a) $T = 8$ (b) 12 m s^{-1}
20 (a) 8 m s^{-1} (b) 900 m
21 (a) $U = 7.67$ (b) 0.782 s
22 (a) 119 m s^{-1} (b) 0.0836 s
23 (a) 1.94 s   (b) 3.10 m s^{-1}
24 (a) 2 m s^{-2} (b) 4 m s^{-1}
25 0.345, 9.65 and 10.3 s
26 (a) 1.75 m s^{-2} (b) $v = 1.75t - 5.5$ (c) $s = 0.875t^2 - 5.5t$ (d) 0 and 6.29 s
27 (a) (ii) 50 s   (b) (i) 644 m   (ii) 78 s
28 (a) 2 m s^{-2} (b) 15 m   (c) Velocity changes suddenly
29 (a) 4 and -4 m s^{-1} (b) 4.44 s

1 (a) $(250\mathbf{i} - 40\mathbf{j})$ m   (b) 253 m (to 3 s.f.)   (c) $(45\mathbf{i} - 14\mathbf{j})$ m s^{-1} (d) 47.1 m s^{-1}
2 (a) $(4\mathbf{i})$ m s^{-1} (b) $(10\mathbf{i} + 6\mathbf{j})$ m s^{-1} (c) $(32.5\mathbf{i} + 12.5\mathbf{j})$ m
3 66.8 m s^{-1}
4 (a) $(4\mathbf{i})$ m   (b) $(34\mathbf{i} + 50\mathbf{j})$ m   (c) 58.3 m
5 (a) $(2t + 1)\mathbf{i} + (t - 2)\mathbf{j}$ (b) $t = 3$
   (c) $(t^2 + t + 30)\mathbf{i} + (0.5t^2 - 2t + 18)\mathbf{j}$ (d) $t = 10$
6 (a) $(2t)\mathbf{i} + 8\mathbf{j}$ (b) $t = 4$
7 (a) $(2\mathbf{i} + 3.1\mathbf{j})$ m s^{-2} (b) 906 m (to 3 s.f.)
8 (a) $(-0.4\mathbf{i} - 4\mathbf{j})$ m s^{-2} (b) $\mathbf{v} = (10 - 0.4t)\mathbf{i} + (8 - 4t)\mathbf{j}$
   (c) 9.2 m s^{-1}
9 (a) 165 m (to 3 s.f.)   (b) $(0.2\mathbf{i} + 0.8\mathbf{j})$ m s^{-2} (c) $(4\mathbf{i} + 16\mathbf{j})$ m s^{-1}, 16.5 m s^{-1}
10 (c) $t = 10$
11 (a) $\mathbf{v} = 0.3t\mathbf{i} - 0.4t\mathbf{j}$ (b) $t = 18$ (c) $\mathbf{r} = (9 + 0.15t^2)\mathbf{i} + (5 - 0.2t^2)\mathbf{j}$ (d) $t = 5$
12 (a) $\mathbf{v} = (2 - 0.1t)\mathbf{i} + (5 - 0.4t)\mathbf{j}$ (b) $t = 12.5$ (c) $t = 20$
13 (a) $(4\mathbf{i} - 4\mathbf{j})$ m s^{-2} (b) $(-8\mathbf{i} + 21\mathbf{j})$ m s^{-1} (c) $\mathbf{v} = (4t - 8)\mathbf{i} + (21 - 4t)\mathbf{j}$ (d) $t = 5.25$

14 (a) $(-2\mathbf{i} - 6\mathbf{j})\,\mathrm{m\,s^{-2}}$ (b) $(4\mathbf{i} + 2\mathbf{j})\,\mathrm{m\,s^{-1}}$ (c) $t = 2$

15 (a) $(0.6\mathbf{i} - 0.8\mathbf{j})\,\mathrm{m\,s^{-2}}$ (b) $18\,\mathrm{m}$

16 (a) $\mathbf{a} = 0.04\mathbf{i} + 0.06\mathbf{j}$ (b) $T = 40$

17 (b) $\mathbf{u} = 5\mathbf{i} + 8\mathbf{j}, \mathbf{a} = -4\mathbf{i} - 6\mathbf{j}$

18 (a) $3.46\,\mathrm{m\,s^{-1}}$ (b) $5.77\,\mathrm{s}$

19 (a) $80.9\,\mathrm{m\,s^{-1}}$ (b) $009°$

20 $122\,\mathrm{m\,s^{-1}}$ on $009.46°$

21 (a) $6.74\,\mathrm{m\,s^{-1}}$ (b) $167.9°$

22 $4.42\,\mathrm{m\,s^{-1}}$ on $118.7°$

23 (a) $2.68\,\mathrm{m\,s^{-1}}$ (b) $63.4°$

24 (a) $\alpha = 31.4°$ (b) $2.21\,\mathrm{m\,s^{-1}}$

25 $3.42\,\mathrm{m\,s^{-1}}$ at $47.0°$ to the downstream bank

26 $113\,\mathrm{m\,s^{-1}}$ on $136.9°$

27 (a) $(-0.3\mathbf{i} + 0.2\mathbf{j})\,\mathrm{m\,s^{-2}}$ (b) $\mathbf{r} = (2t - 0.15t^2 + 20)\mathbf{i} + (-t + 0.1t^2)\mathbf{j}$

    (c) (ii) $5\,\mathrm{m\,s^{-1}}$

28 (b) $\mathbf{r} = (35t - t^2)\mathbf{i} + (45t - 2t^2)\mathbf{j}$ (c) $t = 15$

29 (b) $6.74\,\mathrm{m\,s^{-1}}$

30 (a) $(5\mathbf{i} - 6\mathbf{j})\,\mathrm{m\,s^{-1}}$ (b) (i) $(-5\mathbf{i} + 6\mathbf{j})\,\mathrm{m\,s^{-1}}$ (ii) $10\,\mathrm{m\,s^{-1}}$

Revision exercise 3

1 $255\,\mathrm{N}$

2 $R = 300, F = 250$

3 $600\,\mathrm{N}$

4 (a) $196\,\mathrm{N}$ (b) $176\,\mathrm{N}$

5 (a) $P = 42.3$ (b) $Q = 81.9$ (c) $196\,\mathrm{N}$ (d) $196\,\mathrm{N}$

6 (a) $60 \sin \alpha = 70 \sin 24°$ (b) $28.3°$ (c) $117\,\mathrm{N}$

7 (b) $18.4°$ (c) $F = 9.49$

8 (b) $115\,\mathrm{N}$

9 (a) $W = 58.8\,\mathrm{N}$ (b) $T_1 = 111\,\mathrm{N}$ (c) $T_2 = 94.1\,\mathrm{N}$

10 (a) $\alpha = 13.3°$ (b) $191\,\mathrm{N}$

11 (a) $14\mathbf{i} - 6\mathbf{j}$ (b) $-14\mathbf{i} + 6\mathbf{j}$ (c) $15.2\,\mathrm{N}$ (d) $156.8°$

12 (a) $\mathbf{F}_4 = -27\mathbf{i} - 8\mathbf{j}$ (b) $28.2\,\mathrm{N}$ (c) $163.5°$

13 $T_1 = 113\,\mathrm{N}, W = 80\,\mathrm{N}, m = 8.16\,\mathrm{kg}$

14 (a) $1140\,\mathrm{N}$ (b) $135\,\mathrm{kg}$

15 $266\,\mathrm{N}$ and $342\,\mathrm{N}$

16 (a) $294\,\mathrm{N}$ (b) $176.4\,\mathrm{N}$ (c) (i) 100 (ii) 176.4 (iii) 176.4

    (d) (i) Remains at rest   (ii) Remains at rest   (iii) Accelerates

17 (a) Accelerates   (b) Remains at rest   (c) Accelerates

18 0.417

19 $38.7\,\mathrm{N}$

20 (a) $F = 2.49$ (b) $\alpha = 2.99°$

21 (a) $321\,\mathrm{N}$ (b) No

22 (a) $18.4\,\mathrm{N}$ (b) $12.4\,\mathrm{N}$

23 (a) $22.2\,\mathrm{N}$ (b) $18.6\,\mathrm{N}$

24 (a) $R = 980 - T \sin 30°$ (b) $T = 439$

25 (a) $54.2\,\mathrm{N}$ (b) $13.9\,\mathrm{N}$

26 (a) $92.0\,\mathrm{N}$ (b) $4.52\,\mathrm{N}$ down the slope   (c) 0.0491

27 (a) $\begin{bmatrix} 6 \\ -2.5 \end{bmatrix}$ (b) $6.5\,\mathrm{N}$

28 (c) $0.47\,\mathrm{kg}$

29 (a) $39.2\,\mathrm{N}, 29.4\,\mathrm{N}$ (c) $4.17\,\mathrm{kg}$

Revision exercise 4

1 (a) $1.5\,\mathrm{m\,s^{-2}}$ (b) $1.25\,\mathrm{m\,s^{-2}}$

2 $5600\,\mathrm{N}$

3 (a) $4160\,\mathrm{N}$ (b) $3920\,\mathrm{N}$ (c) $3800\,\mathrm{N}$

4 $0.2\,\mathrm{m\,s^{-2}}$

5 **(a)** 8 N **(b)** 164.8 N
6 585 N
7 0.12 N
8 **(a)** 5.88 N **(b)** $-2.94\,\text{m s}^{-2}$ **(c)** 10.9 m
9 **(a)** $0.4\,\text{m s}^{-2}$ **(b)** 4590 N
10 **(a)** $3.2\,\text{m s}^{-2}$ **(b)** $25\,\text{m s}^{-1}$ **(c)** $50\,\text{m s}^{-1}$
11 **(a)** $0.625\,\text{m s}^{-2}$ **(b)** $0.541\,\text{m s}^{-2}$
12 **(a)** 490 N **(b)** 405 N **(c)** 162 N **(d)** $0.381\,\text{m s}^{-2}$
13 **(a)** 14.9° **(b)** $14.7\,\text{m s}^{-2}$
14 **(a)** $m = 2.5$ **(b)** $f = 3.2$
15 **(a)** $\mathbf{i} + 7\mathbf{j}$ **(b)** $0.2\mathbf{i} + 1.4\mathbf{j}$ **(c)** $1.41\,\text{m s}^{-2}$
16 **(a)** $F = 8$ **(b)** $5.5\mathbf{j}$
17 **(a)** $1.21\,\text{m s}^{-2}$ **(b)** 75.8 N **(c)** $0.260\,\text{m s}^{-2}$
18 **(a)** 5.20 N **(b)** $1.30\,\text{m s}^{-2}$
19 $0.854\,\text{m s}^{-2}$
20 $0.824\,\text{m s}^{-2}$
21 $\alpha = 17.8$
22 **(a)** 37.5 N **(b)** 7.51 N **(c)** $4.80\,\text{m s}^{-2}$
23 **(a)** $\mu = 0.839$ **(b)** $\mu = 0.799$ **(c)** No air resistance
24 **(a)** **(i)** 6750 N **(ii)** $-0.513\,\text{m s}^{-2}$ **(b)** 11 750 N
25 **(a)** 97.6 N **(b)** $4.80\,\text{m s}^{-2}$ **(c)** 31.2 N
26 78.4 N
27 **(b)** **(i)** 0.49 N **(ii)** 0.25
28 **(a)** 196 N **(b)** **(i)** 58.8 N **(ii)** 40 N **(c)** 74.8 N **(d)** 6.12 m
29 **(a)** **(i)** $(2\mathbf{i} - 3\mathbf{j})\,\text{m s}^{-2}$ **(ii)** $(40\mathbf{i} - 60\mathbf{j})\,\text{m s}^{-1}$ **(iii)** $(400\mathbf{i} - 600\mathbf{j})\,\text{m}$ **(b)** 2520 m
30 **(b)** 39.2 N **(c)** 23.7 N **(d)** $1.08\,\text{m s}^{-2}$
31 **(b)** $0.17\,\text{m s}^{-2}$

Revision exercise 5 _____

1 **(a)** $5.88\,\text{m s}^{-2}$ **(b)** 15.7 N
2 **(a)** $3.92\,\text{m s}^{-2}$ **(b)** 41.2 N
3 **(a)** $7\,\text{m s}^{-2}$ **(b)** $1.96\,\text{m s}^{-2}$ **(c)** $1.96\,\text{m s}^{-2}$
4 **(a)** 46.2 N **(b)** 112 N **(c)** 93.3 N
5 **(a)** 28 N **(b)** 5 kg
6 **(a)** $1.5\,\text{m s}^{-2}$ **(b)** 41.5 N **(c)** 3.67 kg

7 **(a)** Light and smooth **(b)** No air resistance **(c)** $\dfrac{g}{49}$ **(d)** $\dfrac{1200mg}{49}$

8 **(a)** $a = \dfrac{(M - m)g}{m + M}$ **(b)** $T = \dfrac{2mMg}{m + M}$

9 **(a)** $\dfrac{g}{5}$ **(b)** $\dfrac{24mg}{25}$

10 **(a)** 11.8 N **(b)** $6.16\,\text{m s}^{-2}$ **(c)** 36.4 N
11 **(a)** $6.3\,\text{m s}^{-2}$ **(b)** 0.398 s **(c)** 31.5 N
12 **(a)** $0.36\,\text{m s}^{-2}$ **(b)** 66.1 N **(c)** 15.4 kg

13 **(a)** $a = \dfrac{mg - \mu Mg}{m + M}$ **(b)** $T = \dfrac{mMg + \mu mMg}{m + M}$

14 **(a)** $1.4\,\text{m s}^{-2}$ **(b)** 42 N **(c)** 33.6 N
15 **(a)** $4\,\text{m s}^{-2}$ **(b)** 40.6 N **(c)** 0.973
16 **(a)** $1.88\,\text{m s}^{-2}$ **(b)** 47.6 N **(c)** 0.415
17 **(a)** $0.556\,\text{m s}^{-2}$ **(b)** 3.56 N **(c)** 1.78 N
18 **(a)** 620 N **(b)** 980 N
19 **(a)** 1370 N **(b)** 1380 N **(c)** 1410 N
20 **(a)** 3300 N, 1520 N **(b)** 5010 N, 2140 N
21 **(a)** $3.26\,\text{m s}^{-2}$ **(b)** 52.2 N **(c)** 0.700 s
22 **(a)** 19.0 N **(b)** 2.44 kg

23 **(a)** 26.7 N **(b)** 6.80 kg
24 **(a)** 49.8° **(b)** 64.8 N
25 **(a)** 3.00 N **(b)** 2.76 m s⁻² **(c)** 21.1 N
26 $a = \dfrac{g(m\sin\alpha - M)}{m + M}$, tension = $mMg\left(\dfrac{\sin\alpha + 1}{m + M}\right)$
27 **(a)** 62.7 N
28 **(a)** Light and smooth **(c)** 28 N
29 **(a)** 0.25 m s⁻² **(b)** 0.559
30 **(a)** 94.1 N **(b)** 3.57 s

Revision exercise 6

1 **(a)** 2.30 s **(b)** 9.45 m
2 **(a)** 10.3 s **(b)** 1130 m
3 **(a)** 0.256 s **(b)** 2.28 m **(c)** 0.080 m
4 **(a)** 6.06 m **(b)** 7.71 m **(c)** 0.933 s **(d)** 6.46 m
5 **(a)** 1.20 s **(b)** 1.54 m **(c)** The ball is a particle and not subject to air resistance.
6 2.36 m
7 **(a)** 0.606 s **(b)** 10.3 m **(c)** 18.0 m s⁻¹
8 **(a)** 3.06 s **(b)** 27.6 m **(c)** 11.5 m
9 **(a)** 1.11 s **(b)** 8.85 m **(c)** 13.5 m s⁻¹
10 **(a)** 53.1 m **(b)** 84.0 m s⁻¹
11 **(a)** 1.25 s **(b)** 28.7 m
12 **(a)** 814 m **(b)** 802 m
13 **(a)** 2.11 s **(b)** 34.1 m
14 **(a)** 2.14 s **(b)** 170 m
15 **(a)** 1.09 s **(b)** 4.63 m
16 7.35 m
17 $V = 14.7$ and $U = \dfrac{20}{3} = 6.67$ (to 3 s.f.)
18 **(a)** 0.217 s and 1.88 s **(b)** 2.66 m and 23.1 m
19 326 m
20 **(a)** 2.46 s **(b)** 19.6 m
21 **(a)** 1.32 s **(b)** 131 m **(c)** 102 m s⁻¹
22 **(a)** 1.01 s **(b)** 23.8 m s⁻¹ **(c)** 25.7 m s⁻¹
23 $\alpha = 49.6°$
24 $V = 29.0$
25 **(a)** $\dfrac{2V}{g}$ **(b)** $\dfrac{2UV}{g}$
26 $\dfrac{2U^2}{g}$
27 **(a)** **(ii)** 35.3 m **(b)** 19.0 m s⁻¹
28 **(a)** Ball is a particle and no air resistance **(b)** **(i)** 0.787 s **(c)** 10.4 m
29 **(a)** **(ii)** 79.5 m **(b)** 76.5
30 **(a)** 1.5 s **(b)** **(i)** −5.7 m s⁻¹ **(ii)** 25.4°

Revision exercise 7

1 1.7 m s⁻¹
2 4 m s⁻¹
3 It travels in the same direction at 0.2 m s⁻¹
4 **(a)** 3 m s⁻¹ **(b)** 1 m s⁻¹
5 6.5 m s⁻¹
6 18.2 tonnes
7 5 m s⁻¹ and 1.5 m s⁻¹
8 24 m s⁻¹
9 6 m s⁻¹
10 8 kg

11 (a) $1.5\,\mathrm{m\,s^{-1}}$ (b) $3\,\mathrm{m\,s^{-1}}$ (c) A will hit B as $v_A = 1$ and $v_B = 0.75$.

12 (a) $10\,\mathrm{m\,s^{-1}}$ (b) No as $v_A = 3$ and $v_{B+C} = 6.67$

13 $(7\mathbf{i} + 4.2\mathbf{j})\,\mathrm{m\,s^{-1}}$

14 $(\mathbf{i} + 4.5\mathbf{j})\,\mathrm{m\,s^{-1}}$

15 $\begin{bmatrix} 6.2 \\ 1.2 \end{bmatrix}\mathrm{m\,s^{-1}}$

16 $\begin{bmatrix} 23 \\ 0 \end{bmatrix}\mathrm{m\,s^{-1}}$

17 $(2.4\mathbf{i} + 3\mathbf{j})\,\mathrm{m\,s^{-1}}$

18 $(4\mathbf{i} + 5\mathbf{j})\,\mathrm{m\,s^{-1}}$

19 (a) $V = \dfrac{2}{7}$ (b) $0.8\,\mathrm{m\,s^{-1}}$

20 $\begin{bmatrix} 3.5 \\ -2.5 \end{bmatrix}$

21 $X = 9.8, Y = 14$

22 $a = -4, b = 2.5$

23 (a) $m = 11$ (a) $2.5\,\mathrm{m\,s^{-1}}$

24 $k = \dfrac{8}{3}$

25 (a) $20\mathbf{j}$ (b) $12\mathbf{i}$ (c) $10.3\,\mathrm{m\,s^{-1}}$

26 $7.5\,\mathrm{m\,s^{-1}}$ and $20\,\mathrm{m\,s^{-1}}$

27 (a) $\begin{bmatrix} 3.5 \\ 6 \end{bmatrix}$ (b) $\begin{bmatrix} -8 \\ -5 \end{bmatrix}$

28 (a) $0.25\,\mathrm{m\,s^{-1}}$ (a) $0.524\,\mathrm{m\,s^{-1}}$

Practice paper 1

1 (a) $15.3\,\mathrm{m\,s^{-1}}$ (b) $1.56\,\mathrm{s}$ (c) $10.7\,\mathrm{m}$

2 (a) $\begin{bmatrix} -1.75 \\ 3 \end{bmatrix}$ (b) $\begin{bmatrix} -3.4 \\ 3.6 \end{bmatrix}$

3 (a) $3.67\,\mathrm{m\,s^{-1}}$ (b) $312°$

4 (a) $79.3\,\mathrm{N}$ (b) 0.727

5 (a) $39.2\,\mathrm{N}$ (b) $P = 9.8\,\mathrm{N}$ (c) (i) $0.754\,\mathrm{m\,s^{-2}}$ (ii) $45.2\,\mathrm{N}$

6 (a) $61.9\,\mathrm{m}$ (b) $7.19\,\mathrm{s}$ (c) $169\,\mathrm{m}$

7 (a) $(\mathbf{i} - 4\mathbf{j})\,\mathrm{N}$ (b) $(2.5\mathbf{i} + 15\mathbf{j})\,\mathrm{m}$

Practice paper 2

1 (a) $1.02\,\mathrm{s}$ (b) $6.60\,\mathrm{m}$ (c) $11.4\,\mathrm{m\,s^{-1}}$

2 (b) $11.5\,\mathrm{N}$

3 (a) $232\,\mathrm{m\,s^{-1}}$ (b) $135°$

4 (a) (i) $4.7\,\mathrm{N}$ (ii) $5.88\,\mathrm{N}$ (b) (i) $0\,\mathrm{m\,s^{-2}}$ (ii) $0.933\,\mathrm{m\,s^{-2}}$

5 (a) $1.96\,\mathrm{m\,s^{-2}}$ (c) Not affected by air resistance.

6 (a) 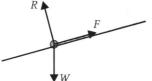   (b) $360\,\mathrm{N}$ (c) $278\,\mathrm{N}$ (d) 0.772

(e) It reduces as the friction force will be less.

7 (b) $62.6\,\mathrm{m}$ (c) $5.48°$ (d) Remains constant   (e) Increases at a constant rate

8 (b) $\mathbf{v} = (0.2t - 2)\mathbf{i} + (3 - 0.4t)\mathbf{j}$ (c) $t = 10$ (d) $t = \dfrac{25}{3}$